Flood Hazard Mapping:

Uncertainty and its Value in the Decision-making Process

A good head and good heart are always a formidable combination.
But when you add to that a literate tongue or pen,
then you have something very special.

Nelson Mandela

Flood Hazard Mapping:

Uncertainty and its Value in the Decision-making Process

DISSERTATION

Submitted in fulfilment of the requirements of
the Board for Doctorates of Delft University of Technology
and of the Academic Board of the UNESCO-IHE
Institute for Water Education
for the Degree of DOCTOR
to be defended in public
on Tuesday, October 4, 2016, at 15:00 hours
in Delft, The Netherlands

by

Micah Mukungu MUKOLWE

Bachelor of Science in Civil Engineering, Makerere University, Kampala, Uganda.
Master of Science in Water Science and Engineering, UNESCO-IHE, Delft,
The Netherlands.

born in Nairobi, Kenya

This dissertation has been approved by the

promotor: Prof. dr. D. P. Solomatine

promotor: Prof. dr. G. Di Baldassarre

Composition of the Doctoral Committee:

Chairman:	Rector Magnificus, TU Delft
Vice-chairman:	Rector, UNESCO-IHE
Prof. dr. D. P. Solomatine	UNESCO-IHE/ TU Delft, promotor
Prof. dr. G. Di Baldassarre	Uppsala University / UNESCO-IHE, promotor

Independent members:	
Prof. dr. ir. H. H. G. Savenije	TU Delft
Prof. dr. P. Bates	University of Bristol
Prof. dr. ir. A. E. Mynett	UNESCO-IHE/ TU Delft
Dr. ir. H. Winsemius	Deltares
Prof. dr. ir. N. C. van de Giesen	TU Delft (reserve member)

CRC Press/Balkema is an imprint of the Taylor & Francis Group, an informa business

Published by:
CRC Press/Balkema
PO Box 11320, 2301 EH Leiden, The Netherlands
Pub.NL@taylorandfrancis.com
www.crcpress.com – www.taylorandfrancis.co.uk
ISBN 978-1-138-03286-6 (Taylor & Francis Group)

Summary

Floods are natural events that can disrupt vulnerable societies and cause significant damages. Floodplain mapping, i.e. the assessment of the areas that can potentially be flooded, can help reduce the negative impact of flood events by supporting the process of landuse planning in areas exposed to flood risk. Flood inundation modelling is one of the most common approaches to develop floodplain maps.

The recent literature has shown that hydraulic modelling of floods is affected by numerous sources of uncertainty that can be reduced (but not eliminated) via calibration and validation. For instance, many studies have shown that models may fail to simulate flood events of magnitude different from that of calibration and validation events. This can be caused by the fact that river flow mechanisms are non-linear and are characterised by thresholds that demarcate flow regimes.

One of the challenges in using uncertain outcomes is that decision makers (e.g. spatial planners) often have to take decisive binary actions, for instance, either to change the landuse (e.g. urbanize) or not. From the perspective of a modeller, one can provide precise (but potentially wrong) results based on both expert knowledge and the results of calibrated-and-validated models. However, this is neither prudent nor pragmatic, given that expert knowledge is variable and unavoidably subjective. As a matter of fact, different modellers using the same input data and models often attain different results. Thus, it is more scientifically sound to provide the results of flood inundation models in probabilistic terms.

The objective of this thesis is to contribute to the scientific work on assessing uncertainty of flood inundation models and develops methods to better support the use of probabilistic flood maps in spatial planning. Thus the impacts of diverse dominant sources of uncertainty (such as input flood hydrograph, model parameters and structure) are assessed by focusing on reduced-complexity models of flood inundation dynamics. Subsequently, novel methods to incorporate uncertain model output in decision making, with respect to spatial planning in floodplain areas, are tested. More specifically, the thesis consists of two main (complementary) parts. The first part deals with the analysis of the major sources of errors in flood inundation modelling, which culminates in the production of probabilistic floodplain maps. The second part shows applications of utility based approaches to aid the decision making processes, when binary decisions are to be made on the basis of uncertain information.

This thesis provides a contribution to the use of probabilistic floodplain maps in decision making, such as spatial planning under flood hazard uncertainty. Using historical hydrological data, 1D, 1D-2D and 2D flood inundation models are used to simulate flooding scenarios. These models are built for two case studies: (i) a mountainous river reach (River Ubaye, France) and (ii) an alluvial river reach (River Po, Italy). Topographic data are derived from frequently used sources of information of different precision and accuracy, namely SRTM (Shuttle Radar Topography Mission), EUDEM (Digital Elevation Model over Europe) and LiDAR (Light Detection and Ranging). In particular, four major components of uncertainty that affect flood modelling outputs are analysed. They include inflow uncertainty (flood discharge derived from a rating curve), parameter uncertainty, model structure and topographic data uncertainty. Input uncertainty was defined in two ways: (i) single segment rating curve parameter uncertainty and (ii) aggregated peak discharge uncertainty components. The boundary condition (inflow hydrograph) uncertainty was found to be considerably more significant than parametric uncertainty. Probabilistic flood hazard maps are generated using a Monte Carlo approach to capture the impact of these sources of uncertainty. Lastly, a new methodology for assessing the benefits of flood hazard mitigation measures (i.e. the KULTURisk framework as a result of an EU FP7 project) was used.

The utility of probabilistic model output is then assessed using two approaches: (i) Value of Information, and (ii) Prospect theory. Implementation of these two approaches is based on the premise of a welfare trajectory, whereby the value of (and generated from) assets and investments in the floodplain accrue over time. Thus, the occurrence of a flood event results in damages that lower the welfare trajectory. Landuse in the floodplain can be altered based on the needs of the community as well as on potential flood risk. In this case, a higher investment yields higher returns, hence, implying a steeper welfare trajectory (and vice versa). A combination of gains of landuse change with a corresponding threat of flood damage (based on a probabilistic floodplain map) exemplifies the spatial planning dilemma that many decision makers have to deal with. In this context, this thesis has demonstrated that probabilistic model outputs can be successfully used to develop flood hazard mitigation strategies and support spatial planning in floodplain areas. Results also point to actual challenges in spatial planning where floodplain locations with higher consequences and uncertainty are identified as requiring additional monitoring.

TABLE OF CONTENTS

LIST OF FIGURES

LIST OF TABLES

Chapter 1

INTRODUCTION

"Start by doing what is necessary,
then what is possible,
and suddenly you are doing the impossible."

Francis of Assisi

1.1 BACKGROUND AND MOTIVATION

Humankind has always had to live and contend with the occurrence of flooding events. More specifically, many societies have settled in floodplains because of their fertile land and transportation accessibility. High population growth rates and consequent human settlements (and investments) in flood prone areas have led to increasing flood risk, which can be seen as a combination of (i) flood hazard and (ii) an exposed vulnerable receptor (Stein and Stein 2014). Moreover, future projections of population growth and climate change suggest that this trend is set to worsen (Winsemius et al. 2015). Hence, supporting flood risk mitigation strategies by improving understanding (and limiting ambiguity) of the spatial distribution of flood risk is of paramount importance.

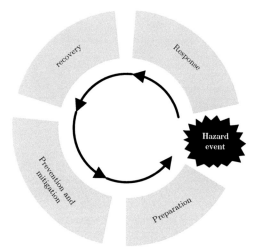

Figure 1.1: Disaster Management cycle (adapted from Lumbroso et al. 2007)

The focus of this thesis is on assessing (and providing methods to cope with) the uncertainty affecting floodplain maps derived by hydraulic models. On a daily basis, decisions are always made under uncertainty, e.g. whether to carry an umbrella or not given a (unavoidably imperfect) precipitation forecast. Uncertainty is part of human existence, and people adjust accordingly based on their understanding, preferences, values and available information. Extending this to flood occurrence (that is partly out of the domain of mans' control), societies have adopted different strategies that can be summarised by the disaster management cycle (Figure 1.1).

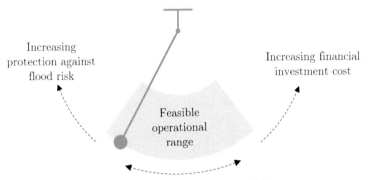

Figure 1.2: Balancing mitigation measures and investment costs

Immediately after the negative impact of a flood event, there is often notable political pressure and higher risk awareness, which then decays over time (Di Baldassarre et al., 2013) and leads to changes in investments (Figure 1.2). Thus, rekindling and maintaining awareness with respect to flood hazards is a key component of flood risk management and this is predominantly covered in the prevention and mitigation part of the disaster management cycle (Figure 1.1). The level of preparedness, the extent and severity of the hazard and the available technical knowhow among several factors largely determine recovery. Focussing on prevention and mitigation can help reduce future flood damages as demonstrated by the EU FP7 KULTURisk project[1] (2011-2014).

The acknowledgment of the inevitability of uncertainty (Koutsoyiannis 2015) is the motivation for this thesis, with a focus on flood inundation modelling and probabilistic flood mapping (Di Baldassarre et al. 2010). Over the past two decades, the research community has gained a better understanding of uncertainty affecting the hydraulic modelling process. These efforts led to the development of several techniques of uncertainty quantification (Montanari 2007, Solomatine and Shrestha 2009, Pappenberger and Beven 2006). Research in flood modelling has also tried to quantify uncertainty and develop methods to communicate it and support the decision making process (Di Baldassarre et al. 2010, Leedal et al. 2010). Meanwhile, other efforts are placed in better understanding of natural systems and improving the

---

[1] An FP7 KULTURisk European Commission funded project (no. 265280) from 2011 to 2013.

mathematical formulation of flood inundation phenomena and their numerical approximation to reduce the uncertainty. In the use of tools (computer models) a draw-back of explicitly providing the uncertainty affecting the model results is that it potentially limits trust in model outputs. Thus, authors such as Refsgaard et al. (2007) and Walker et al. (2003) have called for a stakeholder inclusive participation in the modelling process (from inception to adoption of measures) in order to appreciate and gain trust in the model outputs.

Over the last few decades, there have been improvements in reduced complexity models (Neal et al. 2012b, McMillan and Brasington 2007), numerical schemes (e.g. Bates et al. 2010, Bates and De Roo 2000) and simplification of theoretical conceptual frameworks that simulate flood flows to achieve inundation results within acceptable error bounds (e.g. Dottori and Todini 2011, De Almeida and Bates 2013, De Almeida et al. 2012, Neal et al. 2012a, Bates et al. 2010). Advances in remote sensing and satellite technology have increased the number of topographical data sources for model building. This 'flood of space-borne data' can support flood inundation modelling (Bates 2012, 2004) and create new opportunities to integrate this data into modelling and simulate events in ungauged basins (Di Baldassarre and Uhlenbrook 2012).

With respect to computational concerns, availability of affordable computing power has also opened new avenues for flood inundation modelling. Complex flood flow problems, which were impossible to solve, can now be tackle within a reasonable (and increasingly shorter) duration. Additionally, parallel computing and cloud environments facilitate intense modelling computations (e.g. Glenis et al. 2013, Mukolwe et al. 2015b).

Despite these gains with respect to increased data availability and computer power, there are still challenges regarding the level of accuracy of model outputs. In addition, there are still challenges with respect to the limited number of river gauges. Though, research into advanced measurement techniques and remote sensing (e.g. Hostache et al. 2010, Smith 1997, Schumann et al. 2009) contribute to efforts that address these shortcomings.

The sources of uncertainty can be broadly classified as epistemic and aleatory uncertainty (Van Gelder 2000). Epistemic sources can be reduced by better perceptual (as well as conceptual) understanding of the system and commensurate translation of this conceptual framework into numerical formulations, hence yielding more accurate models. Aleatory uncertainty is an inherent attribute of data used in the modelling frameworks. Counter intuitively, the use of a combination of higher resolution data sources and more complex available modelling does not necessarily yield accurate results (Dottori et al. 2013) given uncertainties in the model structure, evaluation and input data among others. Thus, a balance must be achieved among different factors such as aim of the study, type and accuracy of input (and evaluation) data available, the spatial extent of the study area and the nature of the required outputs.

The inevitability of uncertainty (both aleatory and epistemic) in flood modelling highlights the need for procedural and methodological frameworks to cater for the effects in flood risk mitigation. This thesis work develops an analytical framework to assess uncertainty in flood hazard to support spatial planning within integrated flood risk management. Central to this thesis, is hydrodynamic modelling of floods under uncertainty, followed by an analysis of the potential use of probabilistic floodplain maps in the landuse and spatial planning process. This research work is based on theories of behavioural economics. In particular, new methods are developed by building upon the following theories: (i) Value of Information -VOI (Howard 1966, 1968), (ii) Expected utility (Von Neumann and Morgenstern 1953) and (iii) Prospect theory (Kahneman and Tversky 1979, Tversky and Kahneman 1992).

1.2 RESEARCH OBJECTIVES

The aim of this thesis is twofold: i) to assess the impact of diverse sources of uncertainty in flood inundation modelling, and ii) to use probabilistic floodplain maps, derived from the results of uncertain models, to support the decision making process in flood risk mitigation and spatial planning. The following specific research questions are formulated:

- What are the trade-offs between model complexity, computational efficiency and parameter uncertainty?
- How do input data, model structure and parameters affect the uncertainty of flood inundation models?
- How can probabilistic maps be used by flood risk managers and spatial planners?

1.3 METHODOLOGY

After a review of flood inundation models (Chapter 2), this thesis introduces the case studies: River Ubaye, France and River Po, Italy (Chapter 3). Then, an analysis of the major sources of uncertainty (e.g. boundary conditions, internal roughness model parameters, topographic data and model structure uncertainty) affecting flood inundation models is performed (Chapter 4). In particular, the study focused on the evaluation and quantification of the impact of boundary condition uncertainty and parameter uncertainty in flood inundation models using 1D and 2D model codes of reduced complexity. A Monte Carlo approach based on the Generalised Likelihood Uncertainty Estimation (GLUE) framework is then used to develop probabilistic flood maps and assess flood damage as well as the potential benefits of alternative risk reduction options (Chapter 5).

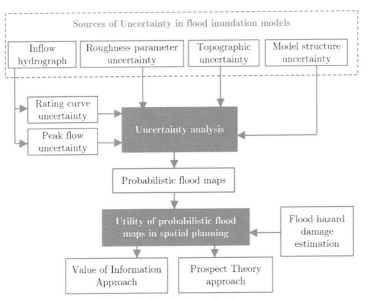

Figure 1.3: Thesis methodology

Subsequently, to explore the value of probabilistic flood maps in the decision-making process, this thesis considers two main theories: (i) Value of Information and (ii) Prospect theory. In particular, the thesis explores the use of model output uncertainty in spatial planning decisions (e.g. either 'changing' the landuse or not under impending flood hazard threat). This is applied to the Ubaye valley case study for which actual landuse changes over a twenty six year period is available. Landuse change consequences with respect to a potential flood hazard are determined using a flood impact assessment (Chapter 5).

1.4 OUTLINE OF THE THESIS

This thesis is composed of two main (complimentary) parts. This first part (Chapters 2 to 4) is about assessing the major sources of uncertainty in flood inundation modelling giving rise to probabilistic floodplain maps. The second part (Chapters 5 and 6) focuses on the value of probabilistic flood maps in the decision making process.

Chapter 2 gives details about current modelling tools for flood hazard assessment. This chapter focuses on flood inundation models and describes the choice of models. It encompasses the perceptual understanding of flood flow dynamics and the mathematical expressions used in flood inundation models. In addition, reasons are given for the suitability of model chosen for this study.

After model selection, Chapter 3 presents model setup information to derive case-specific models and provides a broad description of the cases studies and the available data.

Chapter 4 focuses on uncertainty in flood inundation modelling. This is presented in two sub-parts, (i) an analysis of major sources of uncertainty and (ii) communication of uncertainty to end-users. The chapter concludes with an evaluation of uncertainty in 2D models yielding a probabilistic flood map.

Chapter 5 deals with consequences of flood hazards. This chapter is a precursor to Chapter 6, where floodplain spatial decision making consequences are evaluated with regards to flood damages. The chapter presents a recently developed flood impact assessment framework to derive consequences of flood scenarios.

Chapter 6 focuses on the usefulness of uncertain model outputs (i.e. the probabilistic map derived in chapter 4) in a spatial planning decision making situation. Here, landuse changes are evaluated whereby landuse change decisions have to be made in the floodplain. While consequences of these changes with respect to flood hazard scenarios are evaluated in Chapter 5.

Lastly, a summary of the thesis is presented in Chapter 7, where recommendations and conclusions are also addressed.

Chapter 2

A REVIEW OF FLOOD INUNDATION MODELLING

"Is life so wretched?
Isn't it rather your hands which are too small,
your vision which is muddled?
You are the one who must grow up. "

Dag Hammarskjöld (1905 – 1961)

2.1 INTRODUCTION

Human beings tend to settle in floodplains as they offer favourable conditions for
socio-economic development, e.g. agriculture, access to water, trade, etc. Over the
past decades, population growth and urbanization have triggered greater human
occupation of floodplain areas. An example of this process is depicted in Figure 2.1.
This has contributed to increasing flood losses and fatalities (e.g. Di Baldassarre et
al., 2010).

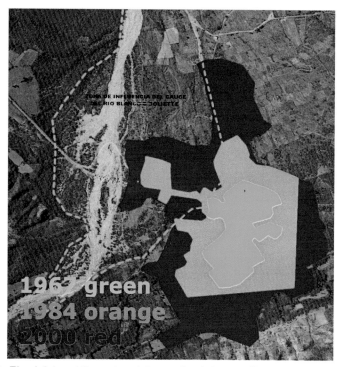

Figure 2.1: Floodplain settlement and future flood damage [Source: Brandimarte et al.
(2009)]

The relationship between humans and floods has a long history (Di Baldassarre et al.,
2013). Experiences with respect to flood hazards are for instance well documented
and exemplified by the nilometers (Popper and Berkeley 1951), early attempts that
Egyptians made thousands years ago to gain a better understanding of the
hydrological regime of the Nile. Additionally, flood marks have been recorded for

many cases for example flood records at Slot Loevestein, the Netherlands (Figure 2.2).

Figure 2.2: Flood marks at Slot Loevestein, the Netherlands [image credit Jan Tilma]
approximate brick width 200mm

These early developments demonstrate attempts to understand and record data related to floods. Modern scientific developments have led to more detailed methods of monitoring flood hazard metrics such as physical scale modelling and computer simulations. Currently, computing power is available in increasingly affordable and portable equipment, as well as development of software specifically tailored for flood risk mitigation.

2.2   FLOOD MODELLING

Prior to accessibility and subsequent popularisation of computing platforms, normative practice was to scale actual river basin physical features down to a laboratory scale model to facilitate testing of different hydraulic scenarios. However, with increasing complexity and demands for more rigorous testing, computational hydraulics became more appealing and practical (Cunge et al. 1980). Scenario analysis for large water related projects is daunting and restrictively expensive to assemble and adjust configurations of physical scale models. Consequently, computer models were popularised due to ease of adaptability and speed at comparatively affordable cost. Importantly, the cost of computers has drastically reduced over the past few decades with a commensurate increase in computing capability and storage capacity. This has availed more opportunities to setup more complex flood inundation models. However, it still is infeasible to achieve outputs of large domain and very complex model configurations within reasonable time frames. Nevertheless, computer models are commonly used to assess as well as acquire knowledge of natural system behaviour, to test hypothesis, and hydrological response scenarios.

Models are approximate representations of reality. Foremost, an aim (either challenge or problem) in nature is identified and an understanding of process involved is then formulated. Subsequently, mathematical interpretation in form of formulae of the conceptual framework is developed. Finally, these equations are then sequenced in a procedural model to obtain a software (Beven 2001, Cunge et al. 1980). Case specific models (e.g. Mukolwe et al. 2015b, Mukolwe et al. 2014, Yan et al. 2013, Md Ali et al. 2015) may then be built using specific data and parameters.

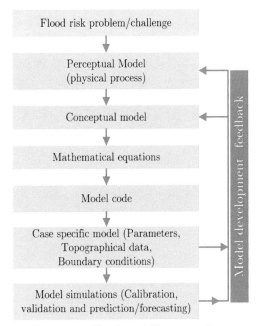

Figure 2.3: Flood modelling procedure

Case specific models are deterministic and consistently give identical results for corresponding set of inputs and parameter sets. However, incongruence between simulated deterministic model space and the actual observed reality space denotes shortcomings of flood modelling tools (Beven 2009). This discordance is the foundation for uncertainty analysis and is discussed further in Chapter 4.

Flood extent estimation can be achieved using tools ranging from simple planar water surface to fully three dimensional flood inundation models (Werner 2004). Hydrodynamic models are available in three major classes 1D, 2D and 3D (Bates 2005). Additionally, there are hybrid model codes that implement combinations of these three dimensions for example 1D-2D (e.g. Prestininzi et al. 2011, Masoero et al. 2013, Bates and De Roo 2000, Domeneghetti et al. 2013).

Despite improvements in available computing resources, modelling of floods using higher dimension model codes is possible but rather time consuming. Additionally, advances in remote sensing have greatly increased the availability of Digital Elevation and Digital Surface Models (DEM and DSM) elevation information (Yan et al. 2015a,

Cobby et al. 2001, Bates 2012). This combination of large amounts of remote sensed data and the need to run models at higher dimensions resulted in the development of reduced complexity flood inundation model codes (e.g. Bates and De Roo 2000). The models (hydraulic models) principally simulate dominant flow processes, thus less significant terms are omitted from core hydraulic water flow equations (McMillan and Brasington 2007, De Almeida and Bates 2013, Neal et al. 2012a, Bates et al. 2010). Simulations are achieved with varying levels of accuracy (e.g. Hunter et al. 2008), thus striking a balance with respect to the level of complexity, computation time, and the inherent data uncertainty. These models have previously been proven to perform within acceptable ranges of accuracy (e.g. Horritt and Bates 2002, Hunter et al. 2008) and are referred to as reduced complexity models in this thesis.

In this chapter, focus is placed on the tools that were used in this study, including model formulation, advantages and reasons for choice of the tools used. The tools are 1D, 1D-2D, and 2D implementations of flood inundation modelling codes, to estimate flood hazard characteristics such as water levels, velocity and flood extent, simulated in an unsteady-flow-state.

2.3 NUMERICAL MODELLING OF FLOODS

A flood can be defined as the inundation of land surface that is usually dry following the exceedance of river flow channel conveyance capacity, damage to the river geometry, or obstruction of water flow (e.g. Apel et al. 2009, Allsop et al. 2007). Flood generation mechanisms and factors may result in floods occurring (i) vertically – upwards, (ii) vertically - downwards and (iii) horizontally (Kundzewicz et al. 2014). Often, floods occur as a result of combinations of these processes. Floods are frequent during periods of increased river discharge caused by intense (long duration) precipitation and (or) rapid snow melt. Additionally, dike breach, debris entrapment, landslide blockage and groundwater rise also increase the intensity of the flood hazard (Di Baldassarre 2012, Flageollet et al. 1996, Marchi et al. 1995).

Perceptual understanding (Figure 2.3) of flooding entails simplifying assumptions that are derived from an understanding of reality. The focus is on riverine flooding due to sustained intense rainfall that causes increased drainage into the main channel. Though vertical (both upwards and downwards) and lateral causal components of flooding are experienced, usually one or more components are more significant (Werner et al. 2005); depending on thresholds such as conveyance capacity, river structures and river channel state. For the cases handled in this thesis, simulated flood events occurred due to intense hydrological conditions upstream of the study areas (Marchi et al. 1996, Flageollet et al. 1996, Flageollet et al. 1999), During a flood, the soil moisture condition is expected to be saturated, hence a high component of predominant overland flow once main channel conveyance capacity is exceeded.

River flood flow is characterised by a low amplitude wave that progressively attenuates downstream due to energy loss. Floods are high magnitude events, thus the longitudinal flow component is dominant. However, once the primary river structure conveyance is exceeded, lateral flows occur as flood water flow fills the floodplains, thus yielding predominant 2D flood flows in the floodplain (Bates 2005). Despite this predominant longitudinal flow, natural river flow is composed of complex flow processes in three dimensions that are amplified at varying sections of the river network such as meanders and bends (Jansen 1979). Flood flow in floodplains is conceptualised as mainly driven by potential differences between water levels in adjacent cells. This understanding of natural flow process facilitates the use of 1D, 2D, and coupled 1D-2D models to estimate flood hazard properties for different river reaches and modelling objectives. The formulation of underlying equations for water flow for classes of models that were used in this thesis is based on conservation of mass and momentum.

In the past, flood routing was commonly executed by use of mainly 1D models, however, with increases in computational power, computing techniques (such as distributed and parallel computing), and development of computationally efficient 2D numerical models, there has been increased use of 2D numerical model approaches to flood mapping (e.g. De Almeida and Bates 2013, Bates et al. 2010, De Almeida et al. 2012, Dottori and Todini 2011, Hunter et al. 2007, Hunter et al. 2008).

2.3.1 Governing flow equations

Flood inundation model setup is heavily data dependant. Ideally, complex models (3D) can be built, however this depends on the amount of data available (for setup and constraining the model parameters). Usually, data required for conditioning and validation is unavailable or difficult to obtain, especially at the same resolution and dimensionality of these complex models. Moreover, computational capacity, with respect to simulation of 3D flows, is prohibitive (especially for Monte Carlo type simulations). Hence, a balance has to be achieved between computational efficiency (computation power) and reality (Bates et al. 2005, FLOODsite 2007).

Hydraulic model equations are derived from the Navier-Stokes momentum equation for an incompressible fluid with a constant density (eq. 2.1).

$$\rho \frac{Du}{Dt} = -\nabla p + \mu \nabla^2 u + F$$

(Schlichting 1979)
eq. 2.1

Where ρ is fluid density, t is time, p is pressure, μ is viscosity and F represents (friction, gravity and coriolis).

$$\nabla u = 0$$

(Schlichting 1979)
eq. 2.2

Combination of Navier-Stokes equation with the continuity equation results in a system of equations that can describe a three dimensional velocity vector, u:

$$u = (u, v, w)$$
eq. 2.3

Where u, v and w are the velocity components in the x, y and z direction of the Cartesian plane respectively.

Complexity of 3D models prohibits setting up of models covering large spatial areas. Moreover, levels of accuracy of available datasets negate the need for increased complexity (Horritt and Bates 2002). Thus, lower dimension and reduced complexity models are more appropriate for use; especially for Monte Carlo type simulations that require a large number of simulations to ensure a robust likelihood values.

2.3.2 HEC-RAS and LISFLOOD-FP Models

The models, 1D - U.S. Corps of Engineers - River Analysis System (HEC-RAS) (Brunner 2010) and 2D (1D-2D) model LISFLOOD-FP (De Almeida and Bates 2013, Bates et al. 2010, Neal et al. 2012a) used in this study, are based on the St. Venant equations for unsteady flow. The flow equations were formulated based on the following assumptions (Cunge et al. 1980):

- One dimensional flow
- Boundary friction and turbulence are accounted for by laws of resistance
- The average channel bed slope is small
- Pressure is hydrostatic given that vertical accelerations are negligible and streamline curvature is small.

The 1D model solves a system of equations including continuity equation and the full St. Venant's 1D momentum equation, where the solution of the partial differential equation is achieved using the Preissmann Numerical Scheme (Preissmann 1961).

$$\text{Momentum:} \quad \frac{\partial u}{\partial t} + u \frac{\partial u}{\partial x} + g \frac{\partial h}{\partial x} + g \left(s - s_f \right) = 0 \qquad \text{eq. 2.4}$$

$$\text{Continuity:} \quad T_w \frac{\partial h}{\partial x} + u \, T_w \frac{\partial h}{\partial x} + A \frac{\partial u}{\partial x} = 0 \qquad \text{eq. 2.5}$$

Where u is velocity, t is time, h is the water-depth, x is a distance, g is gravitational acceleration, s_o is the channel slope, s_o is the friction slope, A is the cross-sectional area and T_{wo} represents top width of flow.

The 2D model (LISFLOOD-FP) code is categorised within a class of reduced complexity models (McMillan and Brasington 2007, Dottori and Todini 2011). Flow within this model is divided into two main parts (i) floodplain flow and (ii) main channel flow (Bates et al. 2010, Neal et al. 2012a). This model solves an inertial approximation of the 1D St. Venant equation, neglecting the convective acceleration term and the derived system of differential equations is solved using a finite difference numerical scheme (De Almeida et al. 2012, De Almeida and Bates 2013). This model was originally built to exploit the emerging remote sensed regular grid elevation information, Digital Elevation Model (DEM) topographic data for 2D models (Bates and De Roo 2000). Consequently flow is de-coupled in the Cartesian x and y directions. Thus, flow fluxes between the topographic raster data cells are calculated in these directions and then water levels in cells updated using conservation of mass.

$$\text{Momentum-x:} \quad \frac{\partial q_x}{\partial t} + gh\frac{\partial(h+z)}{\partial x} + \frac{gn^2|q_x|q_x}{h^{7/3}} = 0 \qquad \text{eq. 2.6}$$

$$\text{Momentum-y:} \quad \frac{\partial q_y}{\partial t} + gh\frac{\partial(h+z)}{\partial y} + \frac{gn^2|q_y|q_y}{h^{7/3}} = 0 \qquad \text{eq. 2.7}$$

$$\text{Continuity:} \quad \frac{\partial h}{\partial t} + \frac{\partial q_x}{\partial x} + \frac{\partial q_y}{\partial y} = 0 \qquad \text{eq. 2.8}$$

Where x and y represent 2D Cartesian directions, t is time, h is the water depth, q_x and q_y are x and y components of unit discharge. An adaptive time-step solution is implemented within the LISFLOOD-FP code (Hunter et al. 2005b).

2.3.3 Why LISFLOOD-FP?

Horritt and Bates (2002) showed that LISFLOOD-FP performed similarly well when compared with TELEMAC-2D (a 2d finite element model code developed by *Electricité de France*) and HEC-RAS (1D) when optimally calibrated. Hunter et al. (2008) present a rigorous benchmarking assessment of commonly used 2D flood modelling codes covering a wide range of formulations in a densely urban area, namely:

- implicit finite-difference solution of full 2D shallow water equations
- explicit finite-difference solution of full 2D shallow water equations
- explicit finite-volume solutions of the full 2D shallow-water equations
- explicit analytical approximations to the 2D diffusion wave equations;
  LISFLOOD-FP model (Hunter et al. 2005b)

This assessment yields acceptable model results with slight differences partially attributed to inertial flow effects in the assessment of inundation extent. However, other sources of uncertainty in this case were micro-topography and continuous slopes along paved surfaces (that magnify inertial effects), and boundary condition inaccuracy (Hunter et al. 2008). Further to this study, developments of the LISFLOOD-FP model code have addressed some of these shortcomings. Bates et al. (2010) presents a quicker model code (compared to earlier diffusive code by Hunter et al. (2005b)) implementing an inertial formulation of the underlying equations of flow. De Almeida and Bates (2013) further assess the applicability of the inertial formulation and show a general acceptable agreement with full-dynamic models for subcritical flows. Thus the model is applicable to river reaches with gentle river channel and floodplain slopes and low amplitude flood waves, which is a common feature for mid- to lower river sections (e.g. Mukolwe et al. 2015b, Yan et al. 2015b). Parallelised versions of the model code (e.g. Neal et al. 2009a) facilitate simulations that can take advantage of now common multi-core architecture computers (Figure 2.4).

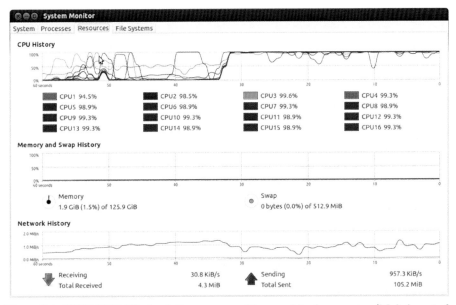

Figure 2.4: **LISFLOOD-FP** simulations on a multi-core virtual computer (Mukolwe et al. 2015b)

For the study Mukolwe et al. (2015b) where several Monte Carlo simulations of 2D models were required, distributed computing was applied on a cloud computing network of virtual computers (Figure 2.5).

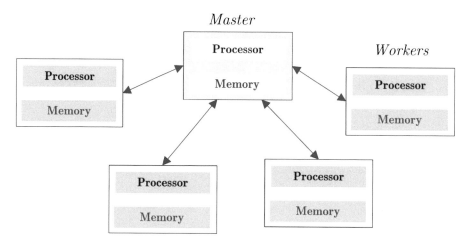

Figure 2.5: Model simulations, Distributed computing on SURF-SARA (https://surfsara.nl) Cloud infrastructure

LISFLOOD-FP is parallelised using 'OpenMP' application programming interface (Neal et al. 2009a) which is optimised to use shared memory, thus making it efficient when the processors and memory are on the same computer. These models are instantiated on multi-processor nodes (workers) and jobs are sent by a master on which controlling scripts are located. Model outputs are then sent back to the master node and synchronised with a local folder. Using freeware/open-source software such as 'Ubuntu Linux' operating system and 'Python' programming module stack (Perez and Granger 2007) to control nodes and write scripts, facilitates up-scaling of the nodes (workers) to limits specified by availability and user capacity.

2.4 Conclusions

The forgoing sections of this chapter have shown the rationale behind the choice of the models used in this case study. Furthermore, given that the work in this thesis was undertaken with the involvement of stakeholders in the KULTURisk project, transparency, simplicity, availability, and robustness of these model codes are considered. To this end, choices are made while basing on previous studies that had shown the suitability of the models. Following model selection, normatively, the next step in model building is to consider data requirements to setup, calibrate, validate and simulate natural phenomena. Thus Chapter 3 focuses on model setup data to derive case specific models.

Chapter 3

CASE STUDIES AND DATA AVAILABILITY

"You cannot carry out fundamental change without a certain amount of madness.
In this case, it comes from nonconformity, the courage to turn your back on the old formulas, the courage to invent the future.
It took the madmen of yesterday for us to be able to act with extreme clarity today.
I want to be one of those madmen. "

Thomas Sankara

3.1 INTRODUCTION

Flood modelling objectives are often determined beyond the control of the modeller. This choice is usually driven by stakeholders interested in acquiring knowledge about a natural system with regards to flood risk mitigation (Walker et al. 2003, Refsgaard et al. 2007). Resource variability and case study peculiarities result in different data collection systems, tools and accessibility issues. In this chapter, data predominantly related to the case studies and flood inundation modelling is addressed.

3.2 CASE STUDY AREAS

In this thesis modelling objectives are based on two case study areas. That is River Ubaye (Ubaye valley), South France and River Po in Italy. These river sections were selected to test components of uncertainty affecting flood modelling. The most significant difference between these two case studies is the river reach length and profile.

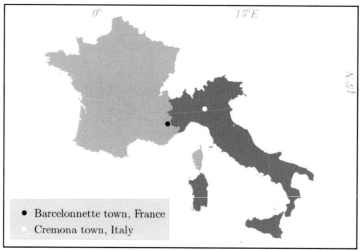

Figure 3.1: Cremona and Barcelonnette towns

River Ubaye study reach (approximately 6 km) is located in the upper reach river profile. While the longer (98km) reach in River Po is in the mid- to lower reach river profile.

3.2.1   River Ubaye, Ubaye Valle (Barcelonnette)

Ubaye valley is located in the French Alps and is a popular tourist destination for alpine related activities. Human settlement in the valley dates back several years, notably was the period of booming trade and migration to Latin America in the 18th century that is evident in cultural Mexican style of construction, still present today. Over time embankments have been constructed, thereby providing settlement areas due to the relative safety of the flood protection; levee effect (Di Baldassarre et al. 2013, Klijn et al. 2004). However, the combination of higher embankments and unknown hydrological conditions can lead to potentially hazardous flooding scenarios.

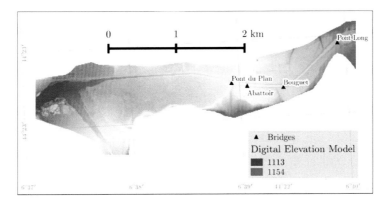

Figure 3.2: Ubaye valley, Barcelonnette town

The valley is confined by steep slopes and is under threat from both hydrological and geomorphological hazards (Thiery et al. 2007). Several steep creeks drain the catchment and a combination of intense mountain-Mediterranean climate and melting alpine snow, results in a rapid river response, hence causing flood flows in the main channel (Flageollet et al. 1996).

Over the years, floodplain settlement has transformed this region into a regular tourist route (and location). Particularly one of the main road arteries connecting France and Italy passes through the area, making it an important hub along this route.

Figure 3.3: Bouguet Bridge over river Ubaye

Several years of flood hazard mitigation have resulted in both structural and non-structural measures. Non-structural measures such as reforestation to retard overland flow have been implemented in the upper parts of the catchment, while structural features such as embankments have been constructed to protect infrastructure from flood hazards. A combination of historical importance of this area and impending threat of flood hazard, with regards to the levee effect, has the potential to cause critical flood damage.

Flooding along Ubaye River can be classified according to the seasonality of occurrence as spring, summer and autumn floods. Summer and autumn floods are mainly caused by intense rainfall and snow melt respectively. Historical hazard records show that spring floods are the highest magnitude hazard events for instance major flood hazard events have been reported in literature in May 1856 and June 1957 categorised as a centennial flood events (Flageollet et al. 1996, RTM 2009). Though, conflicting records add uncertainty to the accuracy related to characteristics of these historical events. Nevertheless, they are categorised as destructive events though accurate actual metrics with respect to accurate peak flow value are unknown.

Figure 3.4: Alpine snowmelt creek Ubaye Valley

Water level measurements that are used in model setup and simulations were obtained from a gauging station (Figure 3.5) located along the reach.

Figure 3.5: River Ubaye, Barcelonnette gauging station

Land planning in France is statutorily regulated. Catastrophic flooding in the Saone and Rhone valleys and south-west France in 1981 led to a law in 1982 that established a disaster (caused by a natural hazard) compensation system. The law further facilitates natural hazard mitigation, compensation and development of localised hazard mitigation plans (strategies). "Barnier" law (Feb, 1995) instituted Risk Prevention Plans (PPR, *Plans de Prévention des Risques*), Figure 3.6.

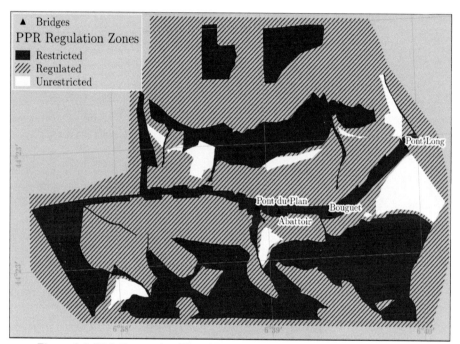

Figure 3.6: Risk Prevention Plan, Ubaye Valley, Barcelonnette (RTM 2006).
Red zone (restricted development), Blue zone (regulated development) and
White Zone (unrestricted development)

These PPR plans are regulatory and are annexed to urban development plans. Plans are enforced by the state through decentralised prefectures (Parisi 2002). The PPR is composed of a presentation note, regulatory zoning and regulations such as, preventive measures, construction and planning rules (e.g. RTM 2009). PPR preparation entails technical and stakeholder participation thus ensuring localised (and prioritised) protection measures as well as strategies. The French spatial planning procedures require public participation in the preparation of risk mitigation plans (Schwarze et al. 2011). In the Ubaye valley, public participation raises risk awareness among the population, hence reduces their vulnerability to flood hazards (Angignard 2011).

In the PPR, multiple natural hazards are accounted for, such as floods, landslides, avalanches, forest fires, earthquakes, volcanic eruptions, and storms. With respect to this range of hazards, methodological guides for each of the hazards are available for inclusion into the PPR. Ubaye valley is mainly at risk of flood and earth movement hazards. Thus the main inputs are Flood prevention plans (PPRI, *Plans de Prevention du Risqué d'Inondations*) and the guide for landslide prevention plans.

French natural hazard mitigation strategy is further boosted by a largely state controlled insurance system that has market-economy elements (Schwarze et al. 2011). This compensation system mitigates the negative effects of natural hazards and also reduces the impact of potential future hazards, thus making the federal government responsible. Following the occurrence of a natural hazard, compensation is triggered by an inter-ministerial decree. The state controls the premiums for all policy holders (Parisi 2002). Legally, a uniform supplement of 12% is collected from all property (Schwarze et al. 2011). Additionally, private insurers are mandated to provide coverage against natural hazards, thus the insurance is mandatorily incorporated into property contracts. Risk is divided into insurable risk and uninsurable risk where uninsurable risk refers to natural hazards and is regulated by the state. The inter-ministerial committee is comprised of members of the home office and the ministry of economy and environment and is tasked with the declaration of a specific event as a disaster. Reinsurance is managed by a state institution; *Caisse Centrale de Réassuarance* (CCR). Private insurers buy subsidized insurance against natural hazards. On the other hand, they may also approach standard reinsurance institutions, though they shall be prone to stringent conditions (Schwarze et al. 2011).

3.2.2   River Po, Italy

River Po is the longest river in Northern Italy and drains an important economic region. It is approximately 650km long, emanating from the Alps and draining into the Adriatic Sea. River flooding along the chosen reach (Cremona - Borgoforte) often results from long duration intense rainfall, thus flood events are common during months around of June and November (Marchi et al. 1996). Years of river training (Castellarin et al. 2010) and subsequent 'levee effect' (Di Baldassarre et al. 2013) have resulted in increased settlements behind the dikes. Historic flooding events have affected human settlement in the floodplains (Masoero et al. 2013, Marchi et al. 1996). In this thesis, discussion is based on flood inundation modelling along an approximately 98km river reach (Cremona - Borgoforte).

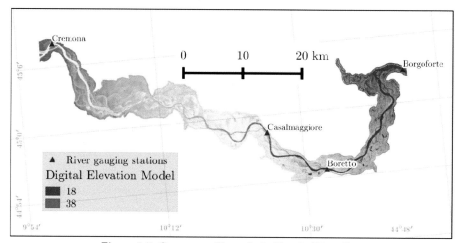

Figure 3.7: Cremona - Borgoforte Reach, River Po

Four gauging stations are located along the reach at Cremona, Casalmaggiore, Boretto and Borgoforte (Figure 3.7 and Figure 3.8) for which water level information is available. Along the study there are reach two water level gauging stations that are used for model conditioning and validation in this thesis.

Figure 3.8: Screenshot online data access portal (AIPO 2015), river Po gauging stations.

Water level data for River Po is collected and maintained by the river basin authority AIPO (*Agenzia Interregionale per il fiume Po* - Interregional Agency for the River Po). These data are automatically collected and transmitted telemetrically. The data can be accessed publically from an online database (Figure 3.8) at a thirty-minute temporal resolution.

3.3   TOPOGRAPHIC DATA

3.3.1   Model geometry input

Case specific flood inundation models are setup with geometric data, inflow discharge data, parametric data and boundary conditions. Frequently, discharge input is sourced from hydrological studies, measured water levels (converted using rating curves) and cascade modelling. Geometric data importantly defines confines of flow controlling features (Farr et al. 2007). With respect to flood inundation modelling, flow controlling features are key components of the model as they cause energy losses (thus attenuation of flow), along with causing complex dynamic flow interactions. Flow controlling features such as embankments (and river banks) determine main channel conveyance capacity, which in turn is the threshold for overtopping and floodplain inundation. Therefore DEM resolution should accurately represent river channel bathymetry and floodplain topography.

Advances in remote sensing data collection and processing, have yielded several sources of topographic information that can be used in model setup (Yan et al. 2015a, Bates 2012, 2004). These datasets are increasingly becoming available either freely or at affordable costs, depending on the type and resolution of the data, with terrain coverage from small to near global scales. Early attempts to build flood models were limited to ground surveys that did not provide dense enough information to adequately represent the main channel and floodplain, thus requiring interpolation (Cunge et al. 1980). For instance Aronica et al. (1998) derived topographic data to build a hydraulic model by digitising topographical maps.

These relatively new datasets are available with varying characteristics (largely dependent on the method of data acquisition) such as accuracy, quality and resolution. LiDAR data is a relatively accurate remote sensed representation of floodplain topography (Cobby et al. 2001). However, due to high acquisition costs, availability of these data is spatially limited. Consequently, lower accuracy (near global coverage) datasets do exist such as SRTM (Shuttle Radar Topography Mission) and ASTER GDEM (Advanced Space borne Thermal Emission and Reflection Radiometer - global digital elevation model). In several instances the data is freely available thus enhancing usage for large scale global studies.

3.3.2   Topographic data sources

Light Detection and Ranging (LiDAR) data is as high accuracy dataset collected by remote sensing. It is usually collected by instruments mounted on low flying aircraft and is characterised by a vertical accuracy of 10-15 cm with an approximate resolution of 1-5 m, hence accurately represents floodplain surface geometry (Bates 2012). Despite high acquisition costs, various national initiatives have collected and availed these data within their territories. For instance the Environmental Agency,

United Kingdom, manages a LiDAR datasets covering approximately 70% of England and Wales. AHN - *Actueel Hoogtebestand Nederland* (Actual Height of the Netherlands) is a Dutch initiative maintaining and distributing surface elevation LiDAR data for the Netherlands (van der Zon 2013). However, for several other locations and regions, LiDAR is largely inaccessible due to restrictive acquisition costs.

Globally, available free to low-cost topographic data offers increasing opportunities for hydraulic modelling of floods. For example, the Shuttle Radar Topography Mission (SRTM) provides the most complete topographic data at a near-global scale. The SRTM elevation product covers areas approximately between 60^0N and 60^0S, about 80% of the Earth Terrain (Farr et al. 2007). SRTM data is available at a resolution of 1 and 3 arc sec (approximately 30m and 90 m respectively).

Due to the data collection technology used for SRTM, radar based interferometric synthetic aperture radar (SAR), the dataset is affected by random noise and radar speckles. Consequently, vertical height accuracy of SRTM topographic data ranges between 5.6 m and 9.0 m (Rodriguez et al. 2006). Previous studies have shown that the absolute height error of SRTM is strongly influenced by topography with vertical large errors in regions with varying relief terrain. On the other hand, in low-to-medium varying terrain areas, vertical errors are lower (e.g. Sanders 2007, Falorni et al. 2005, Wang et al. 2012, Patro et al. 2009). Thus, SRTM is suitable for hydraulic modelling in low relief areas, such as floodplains, rivers (specifically lower and mid-reaches) and river deltas. In this context, a number of scientists have explored the potential of SRTM in supporting large-scale modelling of rivers and floodplains (e.g. Sanders 2007, Neal et al. 2012a, Mersel et al. 2013, LeFavour and Alsdorf 2005).

ASTER (Advanced Space-borne Thermal Emission and Reflection Radiometer) GDEM (Global Digital Elevation Model) is a 30-m spatial resolution DEM, developed using stereo-photogrammetry. Studies have shown that the dataset contains errors that limit its use. Due to reported vertical accuracy of 17 m at 95% confidence level (Tachikawa et al. 2011b, Tachikawa et al. 2011a), this dataset is not commonly used for flood inundation modelling.

EUDEM (European Union Digital Elevation Model) which is a weighted fusion of ASTER-GDEM and SRTM, was released in 2014 by the European Environment Agency (European Environmental Agency 2014). This data is available at a resolution of 1 degree (~30 m), covering several European countries. Considering the aforementioned limitations regarding accuracy of constituent datasets, the potential of EUDEM in flood inundation modelling had not been tested (at the time of writing this thesis). Thus the applicability of the dataset is tested with respect to uncertainty analysis in comparison with SRTM and LiDAR data (Mukolwe et al. 2015b).

3.4   PARAMETRIC DATA

Flood models are not only dependent upon physically measurable quantities but also parameters that represent local (case specific) conditions. In addition, if we consider cascade modelling where one model input is dependent upon other models then the total number of parameters increase. For instance, using the rating curve to generate upstream hydrograph discharges (e.g. Mukolwe et al. 2014, Mukolwe et al. 2015b). Higher dimension models with increased complexity also increase the number of parameters. The challenge is that most often parametric data are not measurable, thus has to be inferred by calibration.

3.4.1   Model parameters

The roughness parameters (Manning's Roughness coefficient) causes energy loss during flow along a river, thus causing attenuation of the flood wave. Despite laboratory scale experiments to characterise these parameters (Chow 1959), accurate representative roughness coefficient are unknown and highly variable from place to place (Beven 2000). Moreover, due to simplified representations of floodplain flow dynamics, roughness values usually compensate for complex flow processes (Romanowicz and Beven 2003). This is best exemplified by taking the example of using lower dimension models to describe flow process, where actual flow processes are more complex with 3D elements of flow. However, if other flow components are not represented then flow dynamics are lumped onto the roughness parameter, therefore. Over-parametersing a model may increase the predictive uncertainty

(Mukolwe et al. 2014), thus, striking a balance between parameterisation, parsimony and available data is necessary.

Roughness values are thus normally estimated by conditioning parameter values. This is done by varying the parameters and comparing simulated model outputs to actual observations. Whereby, the actual parameters used in the model are effective values. This then raises challenges such as parameter non-stationarity and over-parameterisation among others as discussed in Section 4.2.

3.4.2   Inflow discharge

Channel and floodplain flow dynamics are controlled by characteristics of inflow hydrographs such as shape, temporal aspects, peak flow and gradient (of the rising and recession limbs). Discharge hydrographs may be estimated from cascade modelling (e.g. Kayastha 2014, McMillan and Brasington 2008), where a chain of models utilising precipitation data inputs generate discharge hydrographs at river stations. These rainfall runoff models may also be driven by weather generators (e.g. Breinl et al. 2013). However, often, discharge measurements are indirectly quantified from water level information using rating curves (Figure 3.9).

A rating curve is a relationship between waterlevels and corresponding discharge values (for a given river section) which may be described by a polynomial function or piecewise power law (Fenton 2001, Reitan and Petersen-Øverleir 2009, Braca 2008).

$$Q(h) = \begin{cases} 0 & if\ h < h_{0,1} \\ \alpha_1 (h - h_{0,1})^{\beta_1} & if\ h_{0,1} < h \le h_{s,1} \\ \alpha_2 (h - h_{0,2})^{\beta_2} & if\ h_{s,1} < h \le h_{s,2} \\ ... & ... \\ \alpha_m (h - h_{0,m})^{\beta_m} & if\ h > h_{s,m-1} \end{cases} \qquad \text{eq. 3.1}$$

(Reitan and Petersen-Øverleir 2009)

Where Q is the calculated discharge and h is the water level, while α , β and h_0 are parameters.

Typically water levels are measured using a range of measurement equipment such as staffs, floats, pressure and ultrasonic gauges among others (Boiten 2008). Parameterisation of a rating curve is based on a velocity-area measurements to yield a functional form of eq. 3.1.

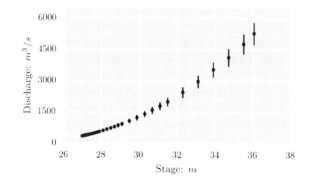

Figure 3.9: Cremona gauging station (River Po) rating curve with 20% and 5cm uncertainties for discharge and waterlevels respectively (Pelletier 1988, Boiten 2008)

With respect to rating curve uncertainties, measurement techniques are done according to best practises (ISO 2010, 1996), however, aleatory uncertainty and changes in river channel properties result in errors that significantly affect hydraulic modelling (e.g. Hunter et al. 2005a). In this thesis, combined effects of rating curve and model parameters are assessed and this is discussed further in Section 4.4.

3.5   CONCLUSIONS

Data sources to support flood inundation modelling are presented, in addition to the location and peculiarities of the case studies that are used in this study. The synthesis of these data sources and effects on flood inundation modelling are addressed in subsequent sections. With the exceptions of floodplain receptor exposure, vulnerability, and economic value data that are presented in Chapter 5. These data are presented within descriptions of the flood damage assessment methodology.

Remote sensed topographical information products particularly do not capture river bathymetry and surfaces below dense vegetation (Schumann et al. 2009), thus, for the study reaches in this thesis, flood plain topography is augmented with river bathymetry data collected from surveys. For the river Po reach, topography DEM is derived by augmenting a 2m DEM of the floodplain from laser scanners mounted on aircraft while multi-beam sonar was used to discretise river bathymetry. Field survey cross-sections (collected by Interregional Authority of the River Po) are used (Di Baldassarre et al., 2009a).

For the Ubaye river study area, floodplain topography LiDAR data collected by sensors mounted on low flying aircraft in 2011. In addition river channel cross-sections collected by RTM are interpolated and fused with the LiDAR data to generate topographic information, while preserving flow controlling features.

Model data requirements facilitate model setup. However, considering data inaccuracies and model structure shortcomings, these models are far from accurate thus, uncertainty in flood inundation models is discussed in the following chapter.

Chapter 4

UNCERTAINTY IN FLOOD MODELLING[2]

[25] So the other disciples told him,

"We have seen the Lord!"

But he said to them,

"Unless I see the nail marks in his hands

and put my finger where the nails were,

and put my hand into his side,

I will not believe."

John 20:25 (The New International Version Bible)

[2] This chapter is based on

Mukolwe, M. M., Yan, K., Di Baldassarre, G. and Solomatine, D. 2015b. Testing new sources of topographic data for flood propagation modelling under structural, parameter and observation uncertainty. *Hydrological Sciences Journal*, 61(9).

Mukolwe, M. M., Di Baldassarre, G., Werner, M. G. F. and Solomatine, D. P. 2014. Flood modelling: parameterisation and inflow uncertainty. *Proceedings of the ICE - Water Management*, 167, 51-60.

4.1 INTRODUCTION

Incongruence between simulated model results and observed variables is the premise for uncertainty analysis in flood inundation modelling. This discordance is potentially perilous especially when outputs are relied upon for decision making in flood risk mitigation, with respect to securing floodplain investment and protecting life. For instance wrong flood warnings and inaccurate hazard mapping will most likely lead to greater damages, loss of life and subsequent loss of credibility of the civil protection institutions. In this respect, it may be disconcerting that current data and state-of-art models and methods, used in flood risk mitigation, are inherently still subject to uncertainties. As discussed in Chapter 3, data for use in model set up is subject to errors depending on methods and tools used for collection. In addition to data uncertainty, this chapter also assesses model structural uncertainty.

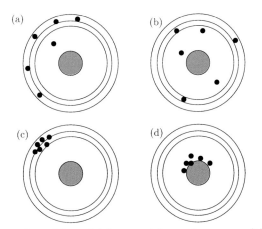

Figure 4.1: Accuracy and precision ; (a) low precision and accuracy, (b) low precision and high accuracy, (c) high precision and low accuracy , (d) high precision and accuracy. Black dots represent model simulation outcomes and red dots represents true values of the phenomenon in focus [adapted from (Streiner and Norman 2006)].

Ideally, model simulations outputs should be both accurate and precise (Figure 4.1 d) where model simulations are as close as possible to the true value (assuming negligible observation uncertainty). For a biased model, we may achieve high precision results that deviate from the true value due to measurement inaccuracies (Figure 4.1 c). On the contrary, model simulation accuracy is difficult to gauge due to

observation uncertainty and model uncertainty. In the following sections of this
chapter, components of uncertainty contributing to occurrence of two extremes Figure
4.1 (a and d) are discussed.

With regards to the inevitability of uncertainty in modelling (Koutsoyiannis 2015),
communication of uncertainty to reduce ambiguity is also presented. Finally, a
probabilistic flood map resulting from an analysis of uncertainty for a real case in
flood modelling is presented.

4.2   UNCERTAINTY ANALYSIS

4.2.1   Introduction

Numerical model output space is deterministic. Thus, for every set of inputs,
parameters and states, unique simulation outputs are consistently obtained. However,
discrepancy emanates from incongruence of the model output space and reality
(Beven 2009). Reality in this case is defined as actual observations of the simulated
variable. Common flood hazard magnitude observation metrics include wrack marks
(e.g. Mignot et al. 2006, Neal et al. 2009b), water levels measurements (e.g. Hunter et
al. 2005a, Mukolwe et al. 2014, Yan et al. 2013) and remote sensed images (Schumann
et al. 2009, e.g. Horritt 2006, Di Baldassarre et al. 2009a). These data are used for
setting up, simulation inputs, constraining model parameters, as well as evaluating
model performance. Data used in flood inundation modelling is channelled from
different sources thus a modeller should be skilful to evaluate the quality of data to
be used. Models built on dis-informative data (Beven and Westerberg 2011) may
potentially increase model predictive uncertainty. According to Boiten (2008)
observation measurements are affected by:

- Observation magnitude
- Number of measured values
- Instrument of acquisition
- Human errors

Few data points and erroneously recorded magnitudes, subsequently affect models when sensitive hydrograph changes are not omitted for instance if peaks are missed. These have profound effect on model performance especially with regards to input uncertainty and consequent effect on simulation outputs.

Table 4.1: Sources of uncertainty assessed in this thesis (adapted from Beven et al. 2014, Neal et al. 2013)

Source	Remarks
Design flood magnitude	Boundary condition uncertainty (Mukolwe et al. 2015a)
Channel conveyance	Three topographical data sources (Mukolwe et al. 2015b)
Rating curve inaccuracy	Rating curve parameter inaccuracy (Mukolwe et al. 2014, Mukolwe et al. 2015b)
Flood plain topography	Three topographical data sources (Mukolwe et al. 2015b)
Model structure	Two model structures; i.e. 1D and 2D (Mukolwe et al. 2015b)
Conditioning data	Not considered
Future changes (landuse and climate)	Not considered
Consequences / vulnerability	Not considered

Universally, flood hazards inherently cause destruction, thus water level gauges hardly work or get destroyed during flood events. It is also common that observation magnitudes surpass measurement ranges of these gauges during floods. On the other hand, advances in remote sensing and satellite altimetry have given rise to more flood data sources (e.g. Yan et al. 2015a, Schumann et al. 2009).

The foundation for uncertainty analysis is characterised by model simulation accuracy and precision (Figure 4.1), with respect to actual observations. Uncertainty analysis entails an analysis of model simulation reliability (Montanari 2007). Statistical (mathematical) methods may be used to analyse the uncertainty. However, using these methods is practically limiting due to numerical and mathematical difficulties in the formulation and epistemic nature of uncertainty (Beven and Binley 2013).

Properties of the residual error structure (model) are unknown, thus, it is appealing to use a Monte Carlo based approach to generate simulations for which error residuals may be analysed. A non probabilistic methodology GLUE (Generalised Likelihood Uncertainty Estimation) is straightforward and commonly used methodology to analyse the model uncertainty (Beven and Binley 2013, Beven and Binley 1992, Beven 2006). Reasons for use of the GLUE methodology stem from the fact that due to imprecision of models, different parameter combinations can simulate model outputs within acceptable limits. This is known as equifinality (Beven 2006). Equifinality aptly illustrates a challenge in flood inundation modelling, where parameter non-stationarity (Romanowicz and Beven 2003) due to uncertainty in model structure and setup data, results in acceptable model simulations (e.g. Figure 4.14). Each model simulation deviates from observations and the level of deviation is determined by a likelihood measure giving a definition of how well the model performed in simulating observed flood metrics. Subsequently, ensemble simulations are weighted by likelihood values. Ideally, the number of bad models is infinite, thus, a criteria of acceptability has to be defined. Criticism of this methodology stems from the subjective choice of likelihood functions (Stedinger et al. 2008) and acceptability thresholds. In this thesis, with respect to simulated water levels, root mean square errors (RMSE) that were greater than the River Po embankment design freeboard (1m), were rejected as non behavioural models.

$$RMSE = \sqrt{\frac{1}{N} \sum_{t=1}^{N} \left(h(t) - \hat{h}(t/\theta) \right)^2} \qquad\qquad \text{eq. 4.1}$$

Where h and $\hat{h}$ represent the observed and simulated water levels respectively at time t. The number of observations is N and θ represents a given parameter set.

A unified assumption in uncertainty analysis methods is that observation error is negligible. Therefore, inference derived from residuals is used to quantify model error. However, observation data accuracy is subject to errors depending upon instrumentation (such as equipment accuracy and malfunctions, magnitude and number of measurements) and data handling errors (Boiten 2008). Thus, the work presented in this thesis makes an assumption that best data collection and measurement techniques were used to collect observation data (e.g. ISO 2010, 1996). GLUE methodology entails sampling of parametric ranges, thus parameter sampling techniques need to be robust to sustain assumptions of stationarity and erodicity

(Montanari 2007). Though, robust sampling of parametric ranges is much simpler for lower dimension models (e.g. Mukolwe et al. 2014, Md Ali et al. 2015) while for higher dimension models, high performance computing environments are more appropriate for intensive model simulations (e.g. Mukolwe et al. 2015b, Neal et al. 2009a).

4.2.2   Methods

Generally, uncertainty in modelling can be classified into (i) aleatory and (ii) epistemological (Van Gelder 2000). Aleatory uncertainty is based on naturally occurring randomness and is an invariable component of data. While epistemological uncertainty is dependent on the state of art knowledge about flood inundation processes (Van Gelder 2000). Acquiring knowledge regarding modelling of flood inundation processes can reduce the epistemological component of uncertainty.

Several uncertainty analysis methods exist, each with particular characteristics regarding mathematical complexity, usability criteria, data dependency and learning curves among others (Pappenberger et al. 2006a, Montanari 2007, Shrestha 2009). Foremost, the most intuitive approach would be to assess properties of the model (Langley 2000). However, this is impractical due to flood inundation model complexity and incomplete knowledge of model errors structures (Montanari 2007). Flood inundation models are characterised by complex non-linearities and model parameter interactions (Romanowicz and Beven 2003). Pappenberger et al. (2006a) outlines challenges lack of uniformity in an understanding of uncertainty terminology and methodology, hence a wiki engine to aid methodology choices is proposed. This endeavour proves to be superior to individual efforts by using widespread collective crowd-sourced experiences to categorise and share knowledge regarding the usability of available methodologies for uncertainty analysis. To this end, methods can be categorised as (i) forward propagation (ii) data dependent (iii) real-time data assimilation and (iv) qualitative methods. This cooperation can foster clearer definitions and procedures hence reduce ambiguity in terms of a shared knowledge base (and experiences) as a code of practice (Montanari 2007, Pappenberger and Beven 2006).

In this thesis, Monte Carlo simulation is adopted where a GLUE (Generalized Likelihood Uncertainty Estimation) approach (Beven and Binley 1992, Beven and Binley 2013). This method overcomes the problem of incomplete knowledge of model error statistical properties by assessing model residuals, hence, evaluating model skill with respect to observations (Refsgaard et al. 2006). The method entails (i) sampling input parameter spaces defined by a-prior knowledge of the parameter distribution (ii) model simulation with random parameter sets. (iii) Accepting models within acceptable range of performance and lastly, (iv) weighing model outputs based on performance. In this case, robustness of these simulations is dependent on the number of model simulations, where output accuracy is inversely proportional to the square root of number of model simulations (Shrestha 2009). As appertains to flood inundation modelling, model computation time is of the essence. Thus, a distributed computing approach (Figure 2.5) is used for Monte Carlo simulations on several virtual computers located on a remote server. This distributed approach is particularly appropriate due to the independence of different parameterized models in the Monte Carlo simulation.

4.3   INFLOW UNCERTAINTY

Statistical methods and models that are used to extrapolate beyond the range of observations need to be robust and accurate (Klemes 1989). Importantly, for flood inundation modelling, rating curves do not accurately account for peak discharges of devastating flood events. Moreover, during these events, measurement equipment is vulnerable to destruction by flood water and debris (e.g. Lecarpentier 1963). In addition, measurements may be affected by equipment error tolerances (e.g. Figure 3.9). During a flood event, cyclic action of sediment erosion and deposition along the river bed results in a highly variable river cross-section during a single event. This causes rating curve inaccuracies and contradicts the usual assumption of river bed stability, especially for steep upper river reaches (Pelletier 1988, Di Baldassarre and Montanari 2009). In addition, extrapolation of rating curve beyond measurement ranges generates erroneous discharge values (Domeneghetti et al. 2012b, Di Baldassarre and Claps 2011).

4.3.1   Rating curve uncertainty

In this study, a single segment rating curve (eq. 4.2) is considered for which two parameters (α and β) are usually determined using linear regression.

$$Q = \begin{cases} 0 & h < H_0 \\ \alpha\left(h - H_0\right)^{\beta} & h > H_0 \end{cases}$$ eq. 4.2

(Herschy 1999)

This rating curve may be reduced to the form:

$$\ln Q = \beta \ln\left(h - H_0\right) + \ln \alpha$$ eq. 4.3

Which is in the form of a straight line with a gradient β and y-axis intercept ($\ln \alpha$). eq. 3.1 may then be parameterised using contemporaneous values of discharge (Q) and water level (h). However, with respect to discharge measurement errors, these parameter values are not certain and will have an amplified effect especially for peak discharges.

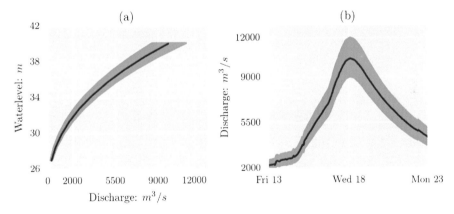

Figure 4.2: **Rating curve parameter uncertainty (a) Cremona rating curve and (b) October 2000 flood event for $\alpha \sim U(4,12)$ and $\beta \sim U(2.3, 3.0)$**

Rating curve errors emanate mainly from changes in the cross-section, river flow properties and discharge measurement techniques (Di Baldassarre and Montanari 2009, Jones 1916, Pappenberger et al. 2006b), namely:

- River bed geometry changes

- Vegetation growth and depletion
- Rating curve appraisal intervals
- Unsteady flow and backwater effects

Multi-segment rating curves (e.g. Reitan and Petersen-Overleir 2009) aptly represent river flow transitions in complex channels as water-levels rise. However, a higher number of segments increase the number of parameters (see eq. 3.1), consequently increasing model predictive uncertainty. In addition, rating curves for river Po river reach do not significantly deviate from the power law (Franchini et al. 1999).

A Monte Carlo approach is applied to assess the impact of rating curve parameters on hydraulic model output. Following a review of historical parameter values for this reach, uniform parameter distributions are used to generate upstream input hydrograph ensembles (Figure 4.2). In addition to rating curve parameters, main channel and floodplain roughness values are also varied based on documented ranges (Chow 1959) and feasible effective values from previous studies (Yan et al. 2013, Di Baldassarre et al. 2009a). The values are main channel roughness from 0.01 to 0.05 $m^{\frac{1}{3}}s$ and for the floodplain roughness 0.04 to 0.11 $m^{\frac{1}{3}}s$. These parameter ranges are further conditioned in a Monte Carlo based approach, while rejecting models with RMSE values greater than a design freeboard of 1m for river Po (Brandimarte and Di Baldassarre 2012). Simulated water levels are compared to observations of a similar high magnitude flood event experienced along the river Po in October 2000. The conditioned model parameter ranges were then tested on another similar magnitude flood event that occurred in November 1994.

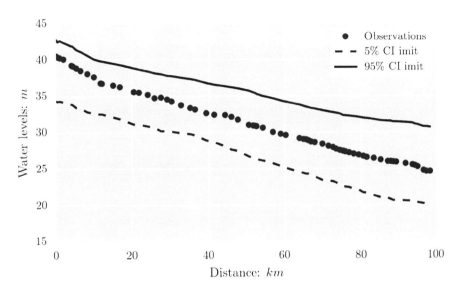

Figure 4.3: High water marks validation October 2000 flood event, HEC-RAS

Additionally, water level simulations of the conditioned models are presented in Figure 4.4.

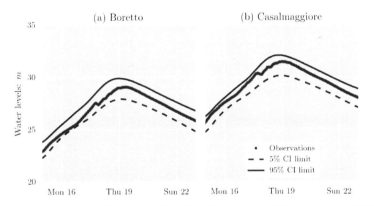

Figure 4.4: Model validation October 2000 flood event (Mukolwe et al. 2014)

With respect to inflow versus parametric uncertainty, models conditioned and validated independently show the dominance of inflow uncertainty over parametric uncertainty, for a 1D model (Mukolwe et al. 2014).

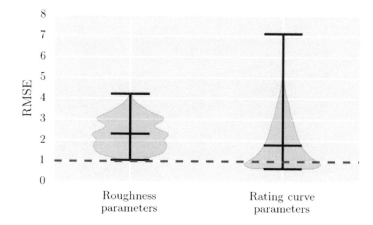

Figure 4.5: Roughness versus rating curve parameter uncertainty, rejection criteria RMSE>1m for October 2000 flood event, river Po

Figure 4.5 shows models for which either roughness or rating curve coefficients were the free parameters. From the figure, it is clear that rating curve uncertainty has a greater effect on model performance. Though there are good models that attain acceptable performance, this wide range of performance also results in very poor model results with RMSE values greater than 7m. This portrays the importance of model input uncertainty with regards to rating curve parameters.

4.3.2   Peak discharge uncertainty

Peak discharge uncertainty is accounted for by generating an ensemble hydrograph for a centennial flood on the Ubaye river similar to a historical flood event in 1957 (Flageollet et al. 1996, RTM 2009), by considering uncertainty in the individual components of uncertainty affecting the discharge (e.g. Di Baldassarre and Montanari 2009). Records show that the flood was characterised as a centennial flood with an approximate peak of 480 m^3/s, a time to peak of 20 hours and an approximate recession duration of 40 hours (RTM 2009). Taking an additive error structure, where Q' indicate the true value of river discharge, Q the observed value and ε the observation error.

$$Q = Q' + \varepsilon \qquad\qquad\qquad \text{eq. 4.4}$$

The observation error is commonly assumed to be a Gaussian random variable with zero mean and standard deviation proportional to the true river discharge and equal to $\beta Q'$ (Di Baldassarre et al. 2011).

Effects of unsteady flow on river discharge data are assumed to be removed in the river discharge data assumed using appropriate methods (Jones 1916, Dottori et al. 2009, Bhattacharya and Solomatine 2005, Fread 1975). Thus, according to Di Baldassarre and Montanari (2009), the global observation error is written as:

$$Q = Q' + \varepsilon + \delta \qquad \qquad \text{eq. 4.5}$$

Where ε denotes the measurement error of the river flow data that are used to build the rating curve, while δ represents the error induced by incorrect rating curve. In particular, under assumptions corresponding to the use of appropriate measurement techniques suggested by ISO (1996), it can be proved that:

$$Q = Q' + \varepsilon + \delta = Q' + \gamma_1 Q'\varepsilon' + \gamma_2 Q'\delta' \qquad \qquad \text{eq. 4.6}$$

Where ε' is Gaussian with zero mean and standard deviation equal to 1, while δ' is a binary variable taking the values +1 or -1 with equal probability. The values of γ_1 and γ_2 are taken from the results of the numerical experiment carried out by Di Baldassarre and Montanari (2009) as 0.027 and 0.384 respectively. And the resulting ensemble hydrographs are presented in Figure 4.6.

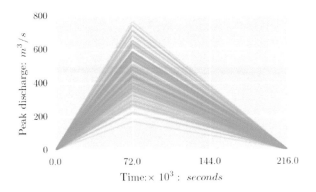

Figure 4.6: Ensemble hydrograph, 1 in 100 year discharge uncertainty

These ensemble hydrographs (Figure 4.6) are used further in Section 4.5.2 to generate a probabilistic flood map.

4.4   MODEL STRUCTURE

Choosing an appropriate flood inundation model is determined by a host of factors such as available data, study objectives, case study characteristics and required outputs among others. It can also be argued that, performance of 1D models are to a certain extent (and confidence level) comparable to other 2D models given uncertainty in data used to build and constrain these models. Moreover, wide scale adoption of 2D models is not applicable to all cases, due to a lack of data to setup and simulate these models.

Results (Figure 4.7 and Figure 4.8) of model calibration (roughness and rating curve parameters) in Mukolwe et al. (2015b), show similarities in model performance for coarse resolution datasets SRTM and EUDEM across two model structures that is, 1D and 2D. However, discrepancies in performance are noted in rising and falling limbs of simulated water-levels. Counter-intuitively 2D models built on the more accurate and precise elevation dataset, LiDAR performed poorly. River Po is heavily embanked and characterised by primary and secondary dikes (Marchi et al. 1996, Marchi et al. 1995, Castellarin et al. 2010, Prestininzi et al. 2011), where predominantly 1D flow is expected during high flows.

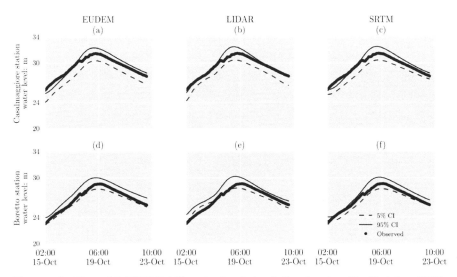

Figure 4.7: 1D model (HEC-RAS) water level simulation for river Po (October 2000), calibration of roughness and rating curve parameters.

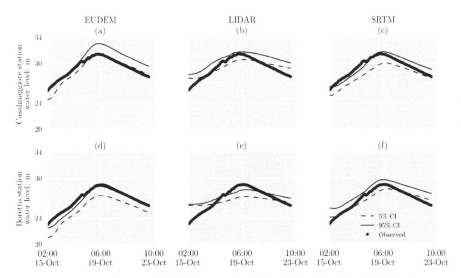

Figure 4.8: 2D model (LISFLOOD-FP) water level simulation for river Po (October 2000), calibration of roughness and rating curve parameters.

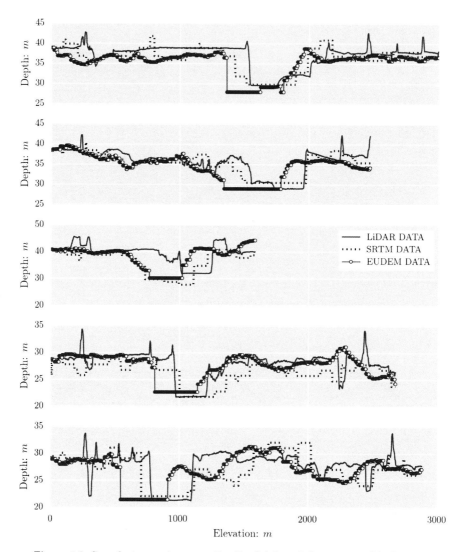

Figure 4.9: Sample transects across the floodplains of the topographic datasets

Flood inundation models are dependent upon flow controlling features such as line elements in model setup elevation data (Werner 2004), thus, a possible explanation for this observed discrepancy in model performance across model structures for LiDAR dataset, is the presence of secondary flow controlling features that create complex turbulent flows that affect model performance (see Figure 4.9).

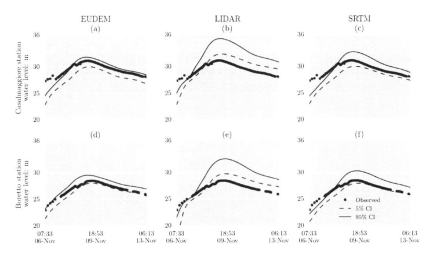

Figure 4.10: 1D model (HEC-RAS) water level simulation for river Po (November 1994),
validation of roughness and rating curve parameters

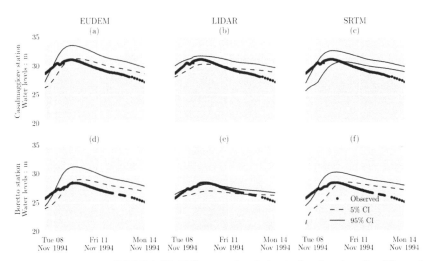

Figure 4.11: 2D model (LISFLOOD-FP) water level simulation for river Po (November
1994), validation of roughness and rating curve parameters

Validation results, Figure 4.10 and Figure 4.11 of models (Figure 4.7 and Figure 4.8)
show poor performance of all 2D models. Models based on coarse resolution dataset
SRTM clearly show a decrease in flood wave celerity and a longer recession limb,
despite hydrograph differences and complexity across events (Bracken 2013), these
flood events used for conditioning and validation, are approximately similar and are

categorised as high magnitude flood events. With varying rating curve parameters, several model input realisations (Figure 4.2b) as well as roughness parameters, does not significantly improve model performance. The main factor in this case is the flood inundation model structure, for which peak water levels and recession limbs were simulated within a 90% confidence interval. Thus, a 1D model that simulates the dominant longitudinal component of flow is preferable for this example.

4.5   COMMUNICATION OF MODEL UNCERTAINTY

Data and model structure uncertainty affects flood inundation modelling outputs and the consequent decisions that are based on these outputs, thus as accurate as possible flood maps are necessary. Hence, there is a need for clear concise information regarding uncertainty in the map production process (Alcock et al. 2010, Montanari 2007, Pappenberger and Beven 2006, Refsgaard et al. 2007, Walker et al. 2003).

Accounting for uncertainty in flood modelling in terms of probabilistic maps (e.g. Leedal et al. 2010, Domeneghetti et al. 2013, Di Baldassarre et al. 2010) and confidence interval ranges (Montanari 2007) is appropriate from a scientific point of view. But, can this probabilistic information be relayed to either stakeholders or general public? The looming fear is that probabilistic information may cause confusion and thus miscommunication (Joslyn and LeClerc 2012, Ramos et al. 2010). Also considering that there are several available uncertainty analysis methods, assumptions, modeller characteristics, models as well as case study properties (Solomatine and Shrestha 2009, Montanari 2007), a question can be raised; what do the uncertain results imply? This is more apparent when dealing with epistemic uncertainties (Beven et al. 2014).

Modelling output end-users require accurate hazard information. On the contrary, outputs are obtained from models conditioned on historical datasets with a level of predictive uncertainty. Hence, assumptions regarding model errors need to be stored and recorded for future assessment (Alcock et al. 2010). Scientific advances in terms of better models and increasing stakeholder participation and awareness, continually applies increasing pressure on flood modellers for clearer more accurate and concise flood hazard (simulation) information (McCarthy et al. 2007, Alcock et al. 2010). Nevertheless, the epistemic component of uncertainty is limited due to the prevailing knowledge. Uncertainty communication therefore is critical to convey information regarding uncertainty in the flood modelling procedure and assumptions made by modellers (Faulkner et al. 2007). Involvement of stakeholders from early stages of flood modelling, especially with regards to assumptions made, effectively improves the understanding of repercussions with respect to model outputs and consequent applicability (Alcock et al. 2010).

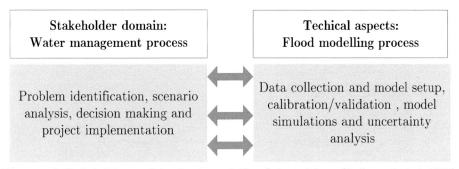

Figure 4.12: Stakeholder participation in modelling (adapted from (Refsgaard et al. 2007)

In this thesis, a stakeholder interaction framework similar to Refsgaard et al. (2007) is followed (Figure 4.13) thus leading to gain a deeper understanding of stakeholder requirements and case study specific peculiarities (Mukolwe et al. 2015a). Modeller-stakeholder interaction (Figure 4.12) facilitates a back-forth review mechanism that updates and refines shared knowledge about case study issues and modelling objectives. Ultimately, the resulting models and outputs are appropriately tailored to the case study. Moreover, stakeholder participation yields a better understanding of the whole process (including residual uncertainty, opportunities and limitations). It is during these interactions that case study peculiarities facilitate the improvement of the flood models including social components regarding calibration of the flood inundation model that is built for the case study area. Information emanating from recorded experiences and photographs give an indication of the expected inundation characteristics (not accurately, albeit indicative of expected model performance). Using this information, a reconstruction of a past flood event in 1957, with little data (Flageollet et al. 1996, Lecarpentier 1963, RTM 2009), is done. Additionally, though the uncertainty of flood damage consequences is not explicitly considered during flood impact assessment (Section 5.2), interactions with stakeholders, improved the understanding of the flood event while evaluating flood hazard and exposure.

Figure 4.13: Stakeholder participation in flood inundation modelling, Ubaye Valley, April 2012 [Photo credit: Dr. L. Alfonso]

In addition to stakeholder-modeller interaction, identified uncertainty related issues that may potentially affect model outputs can be mitigated by having a code of practise to foster guidelines of operation by modellers. This code of practise can effectively reduce ambiguity with regards to uncertainty. The code may contain for instance clearer definitions, language communication channels and formats to improve the interaction and understanding of flood model outputs to stakeholders (Faulkner et al. 2007).

4.5.1   Flood Mapping

Flood risk mitigation strategies are limited and there always exists a residual component of risk. It is impossible to ensure complete protection of human lives and infrastructure from flood hazard damage. Though, the concept of living with floods (resilience) may be considered, having acknowledged that there is inherent uncertainty in components of flood hazard mitigation (Vis et al. 2003). The inability to completely protect societies results in uncertain knowledge of the receptor responses during an extreme flood event (De Bruijn and Klijn 2001). Consequently, the use of flood inundation models while considering significant sources of uncertainty to determine inundation patterns, thereby exhaustively evaluates different possible model simulation outcomes (Di Baldassarre et al. 2010), thus making flood hazard mitigation strategies robust.

A flood hazard map displays information such as the intensity of the flood corresponding to an exceedance probability (Merz et al. 2007). Normatively, flood events peaks and corresponding return periods that are displayed are obtained from a statistical analysis of extremes (e.g. Kidson and Richards 2005). Flood maps can be broadly categorized according to the hazard intensity, the consequence of damage and spatial distribution of the risk (Di Baldassarre et al. 2010).

Table 4.2: Type of flood mapping

Type of Map	Spatial Information contained
Flood danger	Flood danger (lacking exceedance probability)
Flood hazard	Intensity, exceedance probability: for one or more scenarios
Flood vulnerability	Exposure, susceptibility of flood prone elements
Flood damage risk	Expected damage for single/multiple events with exceedance probability

[Source (EXCIMAP 2007, Merz et al. 2007)]

Flood inundation model simulations for a case-specific model, can be used to compare different measures. In addition, using numerical models yields a wide range of temporal and spatially varying output information for instance, water surface profiles, depths, velocities and inundation extent (e.g. Hunter et al. 2005a, Dottori and Todini 2011). A flood map could be described as a map showing the extent of possible flood inundation patterns over an area. Thus, flood maps should contain inundation extent, the magnitude of the inundation depths, and where appropriate, the flow velocity. Furthermore, these maps should give an indication of the potential inhabitants and economic activities that may be affected as well as sensitive installations that may result in pollution of the environment (EU Flood Directive 2007).

During out of bank flow, advancing flood-water is characterized by shallow depths where energy loses are experienced due to the presence of vegetation. The resulting advance may be erratic due to micro-processes (FLOODsite 2007, Merwade et al. 2008). Thus, using a single deterministic binary map for flood inundation mapping is insufficient to fully communicate underlying inherent uncertainty in flood modelling (Beven et al. 2014). To address this, probabilistic maps are generated to account for uncertainty in flood inundation modelling (e.g. Leedal et al. 2010, Horritt 2006, Di Baldassarre et al. 2010, Neal et al. 2013). Uncertainty with regards to flood modelling can be attributed to uncertain peak flow data, data quality, data processing algorithms, extrapolation to rare events and model parameters (Merwade et al. 2008, Merz et al. 2007).

Flood map preparation procedure entails (i) determination of peak discharge, usually following the analysis of observed extreme discharge data. (ii) Propagation of the generated design flood or hydrograph, using a flood inundation model. (iii) And processing of output inundation extent.

For mild undulating floodplains, inundation extent is highly uncertain with respect to a corresponding increase (or decrease) in the water surface elevation. In addition, vertical errors of topographical data contributes to uncertainty (e.g. Mukolwe et al. 2015b). Potential sources of uncertainty that affect flood maps can be attributed to data quality (e.g. Beven and Westerberg 2011), data processing algorithms and extrapolation of data (e.g. Domeneghetti et al. 2012a, Di Baldassarre and Claps 2011). Derivation of a probabilistic flood map is demonstrated in the following section.

4.5.2   Probabilistic flood mapping

Flood inundation models setup (Figure 2.3) consists of calibration and validation steps. Parameterised models, for which immeasurable parameters such as Manning's Roughness coefficient (Chow 1959, Mason et al. 2009, Mason et al. 2003) require either indirect means of determination (e.g. Mason et al. 2009, Mason et al. 2003, Straatsma et al. 2011) or calibration of effective roughness coefficient values (e.g

Gupta et al. 2006, Di Baldassarre et al. 2009b, Götzinger and Bárdossy 2008, Hall 2004, Mukolwe et al. 2014).

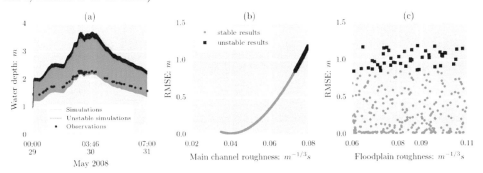

Figure 4.14: 2D model calibration; Main channel roughness sensitivity, 2008 flood event Ubaye river (Mukolwe et al. 2015a)

A 2D flood model of Ubaye valley case study that is conditioned on observed water level data (Figure 4.14) is simulated in a Monte Carlo framework considering input uncertainty derived from an analysis of peak flow uncertainty of a centennial historical flood in 1957 (RTM 2009, Flageollet et al. 1996) and a reconstruction of the hydrograph in Figure 4.6. For this case, each ensemble model output realisation is treated as equally probable, since there was no reliable observation of the actual flood extent.

$$C_i = \frac{\sum_i L_i w_{ij}}{\sum_i L_i}$$

eq. 4.7

(Aronica et al. 2002, Horritt 2006)

Where L_i is the ith simulation likelihood (equal to 1 for each simulation in this case), while w_{ij} is a weight indicating whether the cell is wet or dry. Each simulation was weighted equally (eq. 4.7) to derive a probabilistic map (Figure 4.15).

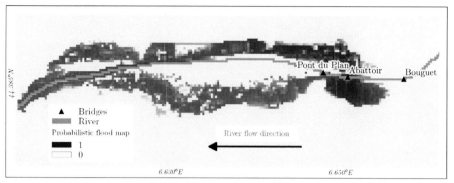

Figure 4.15: Probabilistic map of a 1 in 100 year flood river Ubaye

This derived flood inundation probabilistic map Figure 4.15 contains encapsulated knowledge regarding flooding hazard probabilities and is the basis for testing the applicability of uncertain model output in Chapter 6.

4.6 CONCLUSIONS

Considering the precautionary principle, "When human activities may lead to morally unacceptable harm that is scientifically plausible but uncertain, actions shall be taken to avoid or diminish that harm" (COMEST 2005), flood risk mitigation (in terms of flood hazard mapping) especially under uncertainty is a the responsibility of civil authorities. Uncertain model outputs have to be relied upon during spatial planning (e.g. RTM 2009) for which, future hydrological scenarios and landuse changes are unknown (Beven et al. 2014). This ambiguity caused by uncertainty in flood hazard modelling is undesirable and may lead to negative consequences.

Uncertainty in flood inundation models is related to both aleatory and epistemic components that affects modelling outputs. It is envisaged that clear concise methodologies and statutory guidelines are therefore required to ensure state-of-art practices to limit adverse negative effects of using uncertain information due to suppressed information in deterministic model outputs.

As discussed in this chapter, suppressing and making premature decisions may yield potentially erroneous results in flood modelling. Due to the inevitability of uncertainty in flood modelling, the following sections develop potential frameworks to incorporate uncertain model output (Figure 4.15) and resulting potential consequences of flood hazards in decision making. To this end, Chapter 5 presents a flood risk analysis methodology as a way to evaluate consequences of spatial planning actions and risk prevention (Ronco et al. 2014, Balbi et al. 2012, Giupponi et al. 2015). Chapter 6 presents two methodologies based on the Value of Information (e.g. Bouma et al. 2009) and Prospect theory (Tversky and Kahneman 1992, Kahneman and Tversky 1979) to evaluate the applicability of probabilistic flood maps.

Chapter 5

FLOOD HAZARD MAPS AND DAMAGE[3]

"The task of the modern educator is not to cut down jungles but to irrigate deserts."

C. S. Lewis

---

[3] This Chapter is based on

Mukolwe, M. M., Di Baldassarre, G. and Bogaard, T., 2015a. Chapter 7 - KULTURisk Methodology Application: Ubaye Valley (Barcelonnette, France). *In:* Baldassarre, J. F. S. P. D. ed. *Hydro-Meteorological Hazards, Risks and Disasters.* Boston: Elsevier, 201-211.

5.1   INTRODUCTION

Flood hazard events are not necessarily disasters. Several examples can be cited such as early human settlements in fertile river deltas where farmers benefit periodically from fertile sediment carried by floodwaters (Di Baldassarre 2012). Inundation of uninhabited land does not result in significant damage to people and property due to negligible consequences of flooding.

Flood mitigation measures that address adaptation strategies (to improve receptor resilience) with regards to projected increases in flood hazards extremes are required. Population growth (including conservative) projections indicate that more settlements are expected in floodplains due to pressure on available land. Consequently, prior flood risk mitigation actions such as flood mapping is requisite (e.g. EXCIMAP 2007, Loat and Petrascheck 1997). Statutory requirements, for instance the EU Flood directive (2007), also exemplify the need for prior flood mitigation measures such as flood extent maps.

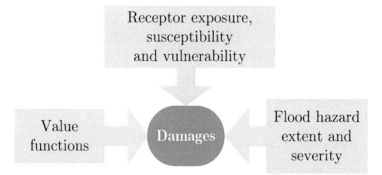

Figure 5.1: Flood damage assessment (Giupponi et al. 2015, Ronco et al. 2014)

Therefore, to aid flood mitigation with respect to spatial planning, an analysis of benefits of risk prevention is required in the form of a flood impact analysis taking into account potential hazard scenarios. This analysis can be summarised as an intersection of hazard and receptor exposure, where impacts are characterised by susceptibility (of the receptors) to the flood hazard. Hazard intensity, receptor susceptibility and vulnerability result in physical damage that can be quantified in economic terms while taking into account repair and reconstruction costs, and willingness to pay cost (e.g. Apel et al. 2004, Jongman et al. 2012, de Moel 2012, Winsemius et al. 2013, Genovese 2006). Within this chapter, the application of a flood impact analysis (Ronco et al. 2014, Giupponi et al. 2015) is presented. The outputs of this analysis are further integrated into proposed methodologies to incorporate uncertain model output into spatial planning decision making frameworks (in Chapter 6). Assessing uncertain hazard information necessitates development and implementation of criteria to combine exposure, susceptibility and value functions, to estimate damages for different scenarios.

The following section presents a detailed step by step implementation of this methodology to the Ubaye valley case study. Furthermore, to illustrate this methodology, a comparison of a baseline and a scenario portraying benefits of flood impact prevention considering structural hazard mitigation measure is presented in the following section. The application of the KULTURisk methodology (Ronco et al. 2014) is presented without consideration of the receptor vulnerability component.

5.2   FLOOD IMPACT ANALYSIS, UBAYE VALLEY, BARCELONNETTE

A key component of the KULTURisk (EU FP7) project, under which this thesis was implemented, was the development of a flood impact analysis framework to evaluate benefits of risk prevention. This methodology is characterised by three main tiers (Figure 5.1) to determine flood damages. Firstly, is the hazard component, which is the physical component of damage by flood water depths and velocities. Secondly, there are actual receptor characteristics that determine receptor exposure and level of physical damage (Ronco et al. 2014). Value functions determined by either a willingness to pay or restoration (reconstruction) costs, are used to evaluate damages to the receptors (Giupponi et al. 2015). Hence, in this thesis damage assessment has been presented in terms of an Economic Regional Risk Assessment (ERRA).

5.2.1   Preliminary analysis

River Ubaye is a fast responding river due to basin shaped characteristics (Section 3.2) while the area of interest around Barcelonnette town has a population of 6851 with a population density of 9 people per square kilometre (INSEE 2014). This area is prone to flood hazards, especially since a large part of river structures have not been changed since the last devastating flood in 1957 (Flageollet et al. 1996). Field visits revealed potential flood mitigation measures that the town of Barcelonnette could implement.

Figure 5.2: Structural flood mitigation measures at Jausiers town, Ubaye valley.

Following the devastation of the 1957 flood event, Jausiers town (upstream of the case study) implemented a series of structural measures (Figure 5.2) that could be potentially implemented downstream such as (i) bridge reconstruction (ii) embankment raising (iii) use of easily replaceable timber bridges, and (iv) having all new construction built at a height of 1.5m above the ground level.

Interaction with stakeholders led to the identification of main channel low conveyance capacity especially at the bridges as being the most important factor that may cause flooding (RTM 2009). Thus in this thesis river conveyance capacity improvement is compared to the baseline scenario.

Table 5.1: KULTURisk methodology application scenarios

Scenario	Description
Baseline Scenario	Current state of the river geometry and structures.
Scenario 1	River channel conveyance enhancement by bridge reconstruction

Having established potential scenarios (Table 5.1), flood hazards corresponding to the structural measures are simulated. An upstream discharge boundary condition corresponding to a 1 in 100 year discharge approximately 480 m^3/s (RTM 2009) is used. These simulations cover two channel configurations corresponding to the current condition (baseline) and scenario 1, where simulations are run with the current main channel state including bridge constriction and same boundary condition without bridges in the model, respectively. Omission of bridge structures from the hydraulic model depict the structural measure involving the enhancement of river channel conveyance.

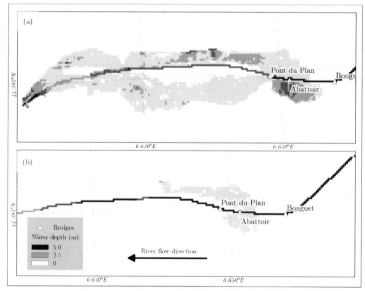

Figure 5.3: Ubaye valley, flood hazard scenario simulations KULTURisk methodology application

Figure 5.3 shows a reduction in the volume of flood water causing inundation. The methodology is limited to receptors inhabiting (situated in) the floodplain, namely: people and economic activities (buildings, road network and agricultural assets). These receptors are identified as being representative with respect to projected flood damages. Two levels of the methodological application (physical and economic impact assessment) are applied while taking into account worst case damage scenarios.

5.2.2   Regional Risk Assessment (RRA)

The impact to people can be expressed as a function of flood hazard, area vulnerability and people vulnerability(DEFRA 2006).

$$Ninj = Nz \times Area\ vulnerability \times People\ vulnerability$$
$$\text{(DEFRA 2006)}$$
$$\text{eq. 5.1}$$

Where *Ninj* is the number of injuries, *Nz* is the number of exposed people at the ground level and basement. The flood hazard rating is the function of the flood depth, velocity and debris factor. While the People vulnerability is a function of the number of very old and disabled or long-term sick people in the floodplain. People vulnerability is the level of susceptibility of the people affected the flood, it is characterised by their ability to respond to hazard. Area vulnerability is defined the function of effectiveness of flood warning, speed of onset of flooding, nature of buildings, it encompasses features of floodplains that affect the safety of the receptors, such as multi-storey buildings where people are safer in higher floors (Wade et al. 2005).

Table 5.2: Area Vulnerability

Parameter	Low impact area	Medium impact area	High impact area
Flood warning	Effective flood warning	Limited flood warning system	No flood warning system
Speed of onset	Long duration lead time (several hours)	Short duration gradual speed of flood onset (an hour or so)	Rapid responding river
Nature of area	Multi-storey apartments	Typical residential area (2 storey homes): commercial and industrial properties	Bungalows, mobile homes, busy roads, single storey schools, camp-sites

Physical impact to receptors is dependent upon stakeholder cultural practices affecting physical exposure to hazard. In this case a worst case scenario was defined where people are located on the ground floor during flood hazard occurrence. Thus, exposure, E is the average house occupancy divided by building floor area.

Figure 5.4: Barcelonnette building layout (Exposure)

According to Giupponi et al. (2015) hazard rating to people (H_{people}) is defined as a function of water depth, velocity and a debris factor.

$$H_{people} = d \times (v + 1.5) + DF$$

eq. 5.2

Where, d is the flood water depth, v is flood water flow velocity (at time of max depth), and DF is the debris factor.

Table 5.3: Debris factor selection (Balbi et al. 2012)

Flood depth (d)	DF
d ≤ 0.25 m	0
0.25 < d < 0.75 m	1
d ≥ 0.75 or v > 2 m/s	1

Thus physical impact to people was calculated as

$$R_1 = 2 \times E \times H_{people} \times \frac{AV}{100} \times SF_{people}$$

eq. 5.3

(DEFRA 2006)

Where R_1 is number of injuries, E is the exposure, H_{people} is the hazard to people, AV is area vulnerability and susceptibility, $SF_{people} = sf_1 + sf_2$, Sf1 (percentage population above 75 years of age) = 9.575 % and Sf_2(percentage terminally ill/disabled people) is estimated as 6% using the national average of disabled people getting government support (INSEE 2014). AV ranges from 3 to 9 representing low to high social vulnerability. A value of 9 is selected indicating a worst case scenario. While for the potential fatalities impact, R_2:

$$R_2 = 2 \times R_1 \times \frac{SF_{people}}{100}$$
(DEFRA 2006)

eq. 5.4

Damage to the road network for this case is defined as inundation consequently requiring routine maintenance to restore the utility. Therefore this is assessed as a percentage of the inundated road network.

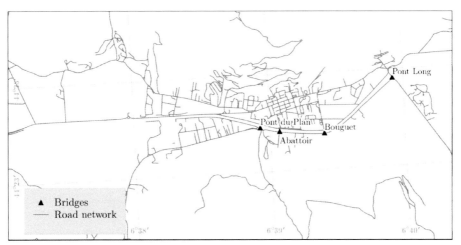

Figure 5.5: Barcelonnette road network

For buildings, impact is assessed by considering the intersection of flood flow velocity and water depth with the corresponding building exposure. The most common type of construction is brick and masonry, thus the building damage assessment by Clausen and Clark (1990) for which constant lines of $v \times d = 7\text{m}^2/\text{s}$ and $v \times d = 3\text{m}^2/\text{s}$ are used to demarcate three classes of damage i.e. total destruction, partial damage, and inundation (Figure 5.6).

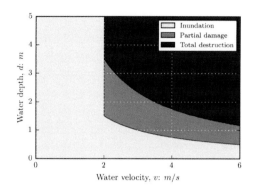

Figure 5.6: Damage criterion for brick and masonry buildings (Clausen and Clark 1990)

Landuse maps of the valley are used to identify agricultural fields (Figure 5.7).

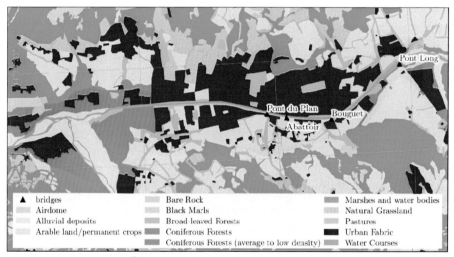

Figure 5.7: Ubaye valley landuse map

Corresponding flood flow metrics (water depth and velocity) are used to characterise the potential damage based on crop damage thresholds.

Table 5.4: Agricultural crop damage thresholds (Citeau 2003)

Agricultural typologies	Spring, summer, and autumn thresholds	
	Maximum depth of flood water (m)	Maximum flow velocity (m/s)
Vegetables		0.25
Vineyards	0.5	0.25
Fruit trees and olive groves	1	0.5

Thus the normalised impact to agricultural crops based on flood metric thresholds (Table 5.4).

Table 5.5: Normalised impact for agricultural crop damage

Agriculture-related risk classes (R_5)	Flood hazard thresholds	Normalised scores
Non inundated	No flood	0
Inundated	Flood metric values below the thresholds	0.6
Destruction of crops	Flood metric values over the threshold	1

5.2.3   Economic - Regional Risk Assessment (E-RRA)

Following the determination of damage factors with respect to physical impacts, an economic assessment is done to evaluate the values of the damages. These damages in monetary terms indicate flood damage consequences of prior spatial planning (or impact prevention) decisions. Damage value factors are related to reconstruction costs, willingness to pay and maintenance costs. Average house rents and construction costs are obtained from INSEE (2014). However, estimates for building content value are estimated in terms of percentages of the building costs. For instance, for residential houses a value of 50% is used (USACE 1996). While damages related to agriculture are based on wheat crop (Messner et al. 2007, Brisson et al. 2010).

Flood damage to infrastructure is assessed as a percentage of inundated road network, thus damage in this case is characterised by minor surface damage, debris deposition where minor road maintenance is required (Doll and van Essen 2008). The following sections elaborate upon detailed equations that are used to determine damage values. The sections are split into two (i) damage to people and (ii) damage to economic activities, such as buildings, infrastructure and agriculture.

(A).   People

Flood damages related to people is calculated as (i) possibility of injury and (ii) potential fatalities. Worth noting is that these are worst case scenarios where maximum exposure is expected. At the core of damage to people is Value of Statistical Life (VSL), which is estimated to be approximately 3.1 Million Euros (OECD 2012).

Costs due to injuries (C_{pi}) are calculated as

$$C_{pi} = E \times R_1 \times B_1 \times VSL \qquad\qquad \text{eq. 5.5}$$

Where R_1 is the rate of injuries (determined in the RRA) and B_1 is the average value of injury compared to loss of life, while damage resulting from fatalities (C_{pf}) is calculated as:

$$C_{pf} = E \times R_2 \times VSL \qquad\qquad \text{eq. 5.6}$$

(B).   Economic activities

Building, infrastructure and agriculture receptors are classified as economic activities. Following RRA assessment of physical impact, value factors were then combined with physical impact. Damage to building was calculated as (i) damage to building structure, which was dependent on structural susceptibility to the flood hazard and (ii) building content damage.

$$D_{sr} = R_3 \times UC_{sr} \qquad\qquad \text{eq. 5.7}$$
$$D_{cr} = R_3 \times UC_{cr} \qquad\qquad \text{eq. 5.8}$$

The main regional road through Ubaye valley passes over the river embankment and a large part of the network is close to the river (Figure 5.5). The resulting damage, D_{rd} is calculated as:

$$D_{rd} = \sum_{i=1}^{nc} [R_4 \times TC]$$ eq. 5.9

Where R_4 is the damage ratio and TC is the clean up and repair cost for the road.

Lastly, agricultural damage (AD) assessment is based on wheat. Taking a yield in terms of tons per hectare (Brisson et al. 2010) and wheat value approximately value per ton, thus damage was calculated as a temporary, one-off loss of agricultural yields (Messner et al. 2007).

$$AD = \sum_{k=1}^{n} [P(k) \times D(k) \times A(k)]$$

(Dutta et al. 2003)   eq. 5.10

Where n is the number of the types of crops, D is the loss per unit area, A is the cultivated area (the exposure of the crop) and P is the cost of the crop per unit area.

5.2.4   Flood damages

Having implemented the preceding KULTURisk damage assessment methodology, flood damages are summed up to derive maps depicting physical risk and damages (Figure 5.8). While the flood damage (consequence) are presented in Figure 5.9.

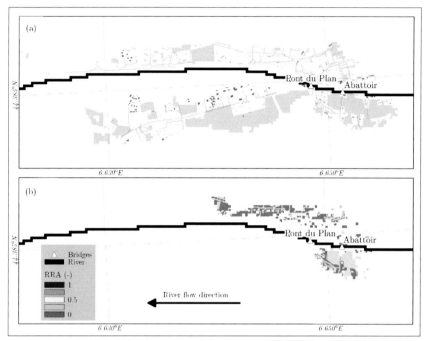

Figure 5.8: Combined receptor RRA maps; KULTRisk methodology

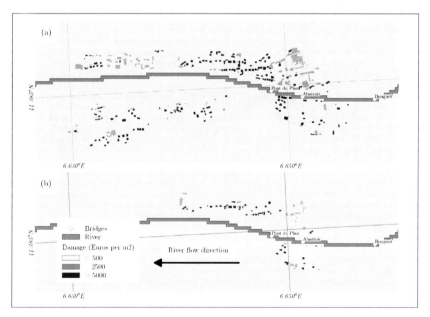

Figure 5.9: Cumulative ERRA maps (a) Baseline (b) Scenario 1; KULTRisk methodology

5.3   UNCERTAINTY IN FLOOD DAMAGE ASSESSMENT

Flood damage is a non-linear function of flood hazard (characteristics) and consequences. Thus interaction of input uncertainties is bound to generate larger output uncertainties. In this thesis, consequence uncertainties (Kreibich et al. 2014), especially with regards to costs involved are neglected. To minimise uncertainties in consequences, analysis in this thesis is based on a comparison of scenarios rather than absolute damage values. Flood hazard uncertainties (as discussed in Chapter 4) are inevitable and are bound to cause epistemological uncertainties.

5.4   CONCLUSION

In this chapter a flood damage assessment resulting from a simulated flood hazard extent is presented. This assessment has been presented as a precursor to Chapter 6 where landuse change decisions consequences are assessed in a similar manner.

Fundamental to flood damage estimation are costs, which can be disintegrated into either direct or indirect and further as either intangible or tangible (Balbi et al. 2013). However, often reported costs are usually incomplete, stating mainly direct costs. Therefore, an integrated cost assessment is requisite to support flood hazard mitigation. This assessment may entail contextualisation, cost assessment, decision support and monitoring as proposed by Kreibich et al. (2014). An exhaustive and relatively complete cost dataset facilitates validation of flood damage assessments, which currently is a challenge for several case studies and was unavailable to validate this assessment (DELTARES 2014).

Chapter 6

Usefulness of Probabilistic flood hazard maps[4]

"The difficulty of tactical manoeuvring
consists of turning the devious into the direct
and misfortune into gain"

Sun Tzu (The Art of War)

---

[4] This chapter is partly based on

Alfonso, L., Mukolwe, M. M. and Di Baldassarre, G. 2016. Probabilistic flood maps to support decision-making: Mapping the Value of Information. *Water Resources Research,* 52(2). DOI: 10.1002/2015WR017378

6.1   INTRODUCTION

The core of this chapter is to address the inevitability of uncertainty in flood inundation modelling (Koutsoyiannis 2015) in the form of probabilistic maps (Krzysztofowicz 2001), which have to be relied upon when making actual spatial planning decisions (e.g. RTM 2006, 2009). To this end, two possible approaches are considered (i) Value of Information and (ii) Prospect theory that may be used to evaluate potential landuse change decisions under the threat of a flood in terms of probabilistic maps to derive implementation strategies.

Landuse changes that occurred in Ubaye valley (Barcelonnette) over a twenty six year period (from 1974 to 2000) are evaluated, while flood damage assessment is evaluated as presented in Chapter 5 with respect to the threat of a centennial flood. With an exception to the workflow in relation to available landuse receptor exposure maps, where building exposure is defined as a coverage of 65% of urban landuse class (e.g. Genovese 2006, Lavalle et al. 2004). The following sections detail these two methodologies with respect to spatial planning and landuse change decisions as applied to the Ubaye valley case study area.

6.2   VALUE OF INFORMATION (VOI)

6.2.1   Introduction

A rational decision maker makes a decision among options based on maximum utility of the decision outcome. Von Neumann and Morgenstern (1953) show that for a given set of probabilities (p) and corresponding outcomes (A), an individual will maximise the expected utility (U) given by

$$U = \sum p_i A_i$$ eq. 6.1

Therefore, a relationship between a probability of flooding (probabilistic flood map Figure 4.15) and corresponding consequences of a flood hazard (Chapter 5) is formulated. Value on information (VOI) can be defined as the maximum value that one is willing to pay for additional information prior to making a decision (Hirshleifer and Riley 1979). Decision making is straightforward for cases where certain information regarding the characteristics of the flood hazard as well as decision makers are available. As exemplified by Alfonso Segura (2010) who presented an example of a person with either an iatrophobic or hypochondriac personality was to seek medical attention. These personality traits are extreme; the hypochondriac will definitely seek medical attention, while the iatrophobic will not seek medical attention. But the challenge then is what if the person is neither a hypochondriac nor iatrophobic? This is a similar scenario that probabilistic maps pose a challenge to decision makers, where the hazard is defined in terms of levels of certitude (Krzysztofowicz 2001), rather than crisp inundation extent boundaries (Beven et al. 2014).

In this case, definite decisions are expected while the hazard is uncertain. Thus VOI is used as a tool to highlight spatial zones where additional information would be required prior to making a decision. Potential spatial planning decisions are assessed under the threat of flooding, in terms of a probabilistic flood map. To achieve this, the following procedure using terminology in Table 6-1 is adopted for a given floodplain location where actions regarding landuse change are considered (floodplain development).

Table 6-1: VOI Terminology

Term	Notation	Description
Location	q	Floodplain area
Action	a_q	Landuse change decision
Message	m_q	Additional information derived from a flood map
State	s_q	Actual state

Thus, the expected utility of an action shall be:

$$u(a_s, p_s) = \sum_s p_s \, u_c(c_{as})$$

eq. 6.2

Where p_s is the perception of flooding at a particular state s out of S possible states, C_{as} is a consequence of action a. Given new information with respect to probabilities of flooding (message, m), the probability vector p_s is updated using Bayes' theorem:

$$p(s/m) = \frac{p(m/s)p_s}{p_m}$$

eq. 6.3

Thus $P(s/m)$ is the updated perception of flooding and $p(m/s)$ is the likelihood of receiving message m given state s. p_m is the probability of message m and is calculated as:

$$p_m = \sum_s p_s p(m/s)$$

eq. 6.4

Therefore, value of information gained is the difference between utilities of actions a_m and a_0.

$$\Delta_m = u(a_m, p(s/m)) - u(a_0, p(s/m))$$

eq. 6.5

Thus Value of Information (VOI) is the expected value of A_m values:

$$VOI = E(\Delta_m) = \sum_m p_m \Delta_m$$

eq. 6.6

6.2.2   Application VOI to Ubaye valley (Barcelonnette)

For Ubaye Valley (Barcelonnette) with respect to a centennial flood threat denoted in probabilistic terms by $P_{(w)i}$ for the ith cell (Figure 4.15), the prior belief vector is:

$$P_{si} = \begin{bmatrix} p(w)_i & (1-p(w)_i) \end{bmatrix}$$

eq. 6.7

If a deterministic map (d) is used, the prior belief vector shall be either $\begin{bmatrix} 1 & 0 \end{bmatrix}$ or $\begin{bmatrix} 0 & 1 \end{bmatrix}$ for wet and dry cells respectively, implying that uncertainty of flooding has either been suppressed or is non-existent. Given new flood hazard information (in form of an actual flood map), the updated perception of flooding is given by:

$$p(m/s) = \begin{bmatrix} R_1 \cap D_1 / (R_1 \cap D_1 + R_1 \cap D_2) & R_1 \cap D_2 / (R_1 \cap D_1 + R_1 \cap D_2) \\ R_0 \cap D_1 / (R_0 \cap D_1 + R_0 \cap D_2) & R_0 \cap D_2 / (R_0 \cap D_1 + R_0 \cap D_2) \end{bmatrix}$$

eq. 6.8

Where $R_{(0,1)}$ is a matrix of binary cells from the flood map and $D_{(1,2)}$ represents landuse change decisions in the floodplain D_1 and D_2 (to either change or retain the landuse respectively) during decision making. Hence, for a perfect flood map, $p(m/s) = \begin{bmatrix} 1 & 0 \\ 0 & 1 \end{bmatrix}$, where flood map cells are congruent with observed flood extent.

In this case, recorded data regarding the observed 1957 centennial flood hazard event is significantly uncertain having been based on an analysis of deposited sediments in the floodplain (Lecarpentier 1963). Based on several studies using LISFLOOD-FP hydraulic flood inundation model, findings show that the model has been successfully validated using precise and accurate measurements of flood extent (Di Baldassarre et al. 2010, Horritt and Bates 2002, Horritt et al. 2007). Therefore, an assumption of a 90% confidence interval to cater for inevitable uncertainty in model simulations is made, thus eq. 6.8 is defined as:

$$P(m/s) = \begin{bmatrix} 0.9 & 0.1 \\ 0.1 & 0.9 \end{bmatrix} \qquad\qquad \text{eq. 6.9}$$

Flooding consequences are used to build a consequence matrix. In this respect, consequences are defined as either damage or gain depending on landuse change decision, corresponding flood hazard damage and lost opportunity costs. This creates a set of options corresponding to different landuse change scenarios for which utilities are calculated. Consequences can be defined as resulting states, due to flooding, based on either change $(D2)$ or no-change $(D1)$ decisions. Using a welfare trajectory (Green et al. 2011) for floodplain receptors Figure 6.1 is drawn.

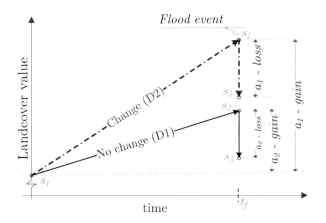

Figure 6.1: Landuse change decision consequences (adapted from (Green et al. 2011))

Consequences of landuse change decisions (related to the effect of a flood) result from appreciation; that is s_1-s_4 and s_1-s_2, and flood damages are s_2-s_3 and s_4-s_5. Higher returns for investment are visible in slope s_1-s_4 being greater than s_1-s_2. Consequently, higher losses are expected for areas in which an investment was made, thus s_4-s_5 is greater than s_2-s_3. This intuitively relates to real case scenarios where flood hazard in built-up areas cause considerable damage as compared to rural locations. The consequence matrix is summarised in Table 6-2 while taking a 'do nothing' case of no-change and resulting damage (s_3) as the baseline to evaluate relative damages.

Table 6-2: Consequence matrix (C_{as})

	D1 (no landuse change)	D2 (land cover change)
Flood	$c_1 = s_3 \text{-} s_3 = 0$	$c_2 = s_3 \text{-} s_5$
No-flood	$c_3 = s_3 \text{-} s_2$	$c_4 = s_4 \text{-} s_3$

Decision consequences are based on landuse changes between the years 1974 and 2000. During this period decisions made (with regards to spatial planning) are determined by evaluating prior and post state of the landuse map pixels, and resulting consequences in terms of flood hazard damage are evaluated using the KULTURisk methodology (Chapter 5). The welfare slope trajectory is evaluated as an appreciation of landuse value at an annual interest rate of 3% (INSEE 2014).

$$P_t = P_0 \left(1 + i\right)^n \qquad\qquad \text{eq. 6.10}$$

Where P_t is landuse value at time t corresponding to occurrence of a flood hazard, while P_0 is the initial landuse pixel value when decisions are made. Value appreciation is based on an annual rate of i% for n years.

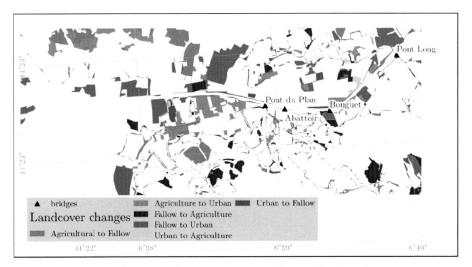

Figure 6.2: Landcover changes 1974 to 2000 Ubaye valley, Barcelonnette

To construct a consequence matrix (Table 6-2), four possible achievable states depending on the decision that is; s_2, s_3, s_4 and s_5 are evaluated. These states correspond to *'not flooded-not changed'*, *'flooded-not changed'*, *'not flooded-changed'* and *'flooded-changed'* from Figure 6.1.

The calculation of VOI is illustrated by randomly selecting a floodplain location from Figure 6.2 . The point is characterised by a probability of flooding $p(w)_i = 0.794$ and a consequence matrix as follows; $C_{as} = \begin{bmatrix} c_1 & c_2 \\ c_3 & c_4 \end{bmatrix} = \begin{bmatrix} 0 & -220.757 \\ -30.579 & 218.485 \end{bmatrix}$

The prior belief vector for the selected point is:

$$P_{s4} = \begin{bmatrix} 0.794 & 0.206 \end{bmatrix}$$

Hence from eq. 6.2

$$u(a_{dc}, p_s) = 0 \times 0.794 + (-30.579) \times 0.206 = -6.299$$

$$u(a_c, p_s) = (-220.757) \times 0.794 + (218.485) \times 0.206 = -130.273$$

The indices c and dc refer to the "*change*" and "*no change*" decisions respectively. The maximum utility of the two actions is:

$$\max(-6.299, -130.273) = -6.299$$

Considering the uncertainty of the flood inundation model the revised beliefs from eq. 6.4 are

$$p_s^c = 0.9 \times 0.794 + 0.1 \times 0.206 = 0.735$$

$$p_s^{dc} = 0.1 \times 0.794 + 0.9 \times 0.206 = 0.265$$

Thus

$$p_m = \begin{bmatrix} 0.735 & 0.265 \end{bmatrix}$$

And the revised beliefs are:

$$p(s/m) = \begin{bmatrix} \left(\dfrac{0.9 \times 0.794}{0.735}\right) & \left(\dfrac{0.1 \times 0.794}{0.265}\right) \\ \left(\dfrac{0.1 \times 0.206}{0.735}\right) & \left(\dfrac{0.9 \times 0.206}{0.265}\right) \end{bmatrix} = \begin{bmatrix} 0.972 & 0.300 \\ 0.028 & 0.700 \end{bmatrix}$$

The expected utility after accounting for the flood model uncertainty and using eq. 6.3:

$$u\big(a_{dc}, p(s/m_{dc})\big) = 0 \times 0.972 + (-30.579) \times 0.028 = -0.856$$
$$u\big(a_{dc}, p(s/m_c)\big) = (0) \times 0.300 + (-30.579) \times 0.700 = -21.405$$
$$u\big(a_c, p(s/m_{dc})\big) = (-220.757) \times 0.972 + (218.485) \times 0.028 = -208.458$$
$$u\big(a_c, p(s/m_c)\big) = (-220.757) \times 0.300 + (218.485) \times 0.700 = 86.712$$

The maximum utilities for the actions "*change*" and "*no change*" are:

$$u^c_{max} = \max(-208.458, 86.712) = 86.712$$
$$u^{dc}_{max} = \max(-0.856, -21.405) = -0.856$$

The information gained is calculated from eq. 6.5:

$$\Delta_c = 86.712 - (-6.299) = 93.011$$
$$\Delta_{dc} = -0.856 - (-6.299) = 5.443$$

Therefore, the VOI for the selected floodplain location is calculated using eq. 6.6

$$VOI = 0.265 \times (93.011) + 0.735 \times (5.443) = 28.648$$

The VOI methodology is then applied to the whole floodplain to yield a VOI map in Figure 6.3.

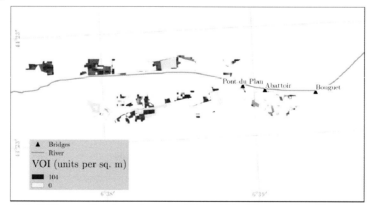

Figure 6.3: VOI map; Landuse changes 1974 to 2000 Ubaye valley (Barcelonnette)

From Figure 6.3, areas characterised by high VOI of information correspond to floodplain locations that have considerable investment by changing the landuse from fallow to urban (Figure 6.2). These locations also correspond to flood flow paths where the potential flood damage is high. Furthermore, these locations with high VOI values correspond to Blue zones in the PPR of the area (Figure 3.6) where regulation is required prior to development.

Further analysis of the VOI values, probabilistic map and landuse changes are presented in Figure 6.4 and Figure 6.5. For this assessment, VOI values are classified in 5 classes ranging from minimum to maximum, specifically (1.899, 22.2], (22.2, 42.4], (42.4, 62.6], (62.6, 82.8] and (82.8, 103]. The first, middle and last classes are labelled as Low, Medium and High VOI.

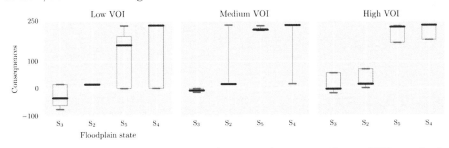

Figure 6.4: Flood damage scenarios (Figure 6.1) corresponding to VOI magnitudes

Figure 6.4 shows that as low VOI is characterised by several combinations of flooding (and damage) scenarios, however uncertainty is considerable lower for high VOI. The figure also shows that rationally it is always necessary to change the landuse class to higher value. Additionally, a similar assessment of corresponding probabilities is also presented in Figure 6.5.

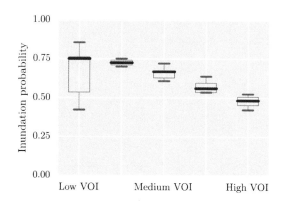

Figure 6.5: Probabilistic map values corresponding to VOI magnitudes

Clearly, as VOI of information increases so does the likelihood of flooding. These likelihoods decrease towards 0.5. This tendency, implies that at maximum uncertainty, likelihood approximately 0.5, either decision is probable.

6.3   PROSPECT THEORY

6.3.1   Introduction

Expected utility does not address peculiarities of decision makers that affect actual decision making. Thus, non-expected utility theorems have been developed to address these shortcomings (Starmer 2000). For example, the Prospect theory (Kahneman and Tversky 1979, Tversky and Kahneman 1992), which is a common alternative to expected utility theorems, and defines how people actually make choices from among alternatives, for which probabilities of occurrence are known. It accounts for the risk appetite of the decision maker as well as loss aversion, whereby, losses have a greater effect than gains of similar magnitude. In addition, diminishing sensitivity of the value function is also addressed where relative changes in lower values (of loss and gain), have a greater effect than in higher values. These characteristics of Prospect theory are relevant to spatial planning (with respect to flood hazard mitigation), where strong element of risk aversion is portrayed by negative effects of loss aversion. Loss of human life (due to a flood hazard) is a significantly negative outcome that erodes the confidence in authorities charged with civil protection, especially for elected administrative officials. The value function has the following form:

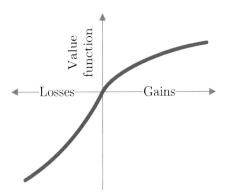

Figure 6.6: Prospect theory; theoretical value function

From Figure 6.6 the value function (v) can thus be defined as:

$$v(x) = \begin{cases} x^{\alpha} & \text{if } x \geq 0 \\ -\lambda(-x)^{\beta} & \text{if } x < 0 \end{cases} \qquad \text{eq. 6.11}$$

Where x is a consequence of an option to be selected, while α, β and λ are factors that denote characteristics of decision makers. λ denotes loss aversion for which values greater than unity imply than individuals are more sensitive to losses than gains. Loss aversion aptly reflects flood hazard mitigation efforts whereby civil protection is geared towards minimising loss of life and damage to property. Losses and damage are undesirable thus cause considerable political pressure for solutions. Values of α and β that are less than one imply risk aversion and risk seeking with respect to the gains and losses respectively (Neilson and Stowe 2002).

It has also been shown that decision makers are inclined to under weigh large probabilities and overweigh lower probabilities. Thus the probability weighting functions w^+ and w^- (positive and negative respectively) are:

$$w^+(p) = \frac{p^{\gamma}}{\left(p^{\gamma} + (1-p)^{\gamma}\right)^{\frac{1}{\gamma}}}, \; w^-(p) = \frac{p^{\delta}}{\left(p^{\delta} + (1-p)^{\delta}\right)^{\frac{1}{\delta}}} \qquad \text{eq. 6.12}$$

Where p is the probability of an outcome and parameters γ and δ are shaping parameters. Consequently, computed cumulative utility (CU), derived by weighting of probabilities and relative outcomes, according to Tversky and Kahneman (1992) is:

$$CU = \sum w\big(p(x)\big)v\big(x\big)$$ eq. 6.13

With respect to actual flood mitigation, where decisions are made with uncertain flood hazard information, varying stakeholder characteristics negate the use of a single formulation to evaluate decisions. Thus this theory facilitates parameter adjustments to reflect varying priorities and decision maker characteristics across case studies.

6.3.2   Making a decision

The decisions evaluated are similar to VOI in Section 6.2. However, the difference is that consequences in terms of landuse changes are relative to values. This is more direct and intuitive. Thus, making a decision to urbanise an area given a probabilistic flood map can be visualised in form of the resulting prospects for different outcomes in terms of a flood hazard event.

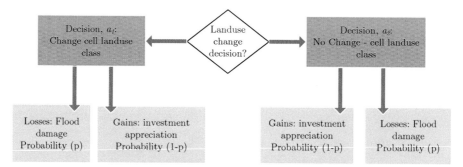

Figure 6.7: Decision making prospects (see also Figure 6.1)

6.3.3   Prospect theory application to Ubaye valley (Barcelonnette)

Parameters from eq. 6.11 and eq. 6.12 define stakeholder characteristics. In terms of spatial planning, using uncertain hazard information different stakeholders will take varying risks and will be prone to different levels of loss aversion that is based on localised factors.

In this thesis, two examples of stakeholders based on literature are chosen. However, further opportunities to derive stakeholder characteristics with respect to flood inundation modelling and spatial planning may be undertaken. Generic parameter values are used from the original Prospect theory paper (Tversky and Kahneman 1992). In addition, for comparison purposes, an additional set of parameters derived from several studies as summarised by Booij et al. (2010) are also evaluated.

Table 6-3: CPT Coefficients; Where TK92 refers to Tversky and Kahneman (1992) and B10 refers to Booij et al. (2010)

	Description	Symbol	Parameters	
			TK92	*B10*
1	Power coefficient, gains	α	0.88	0.69
2	Power coefficient, losses	β	0.88	0.86
3	Loss aversion	λ	2.25	2.07
4	Probability weight, gains	γ	0.61	0.69
5	Probability weight, losses	δ	0.69	0.72

Thus resulting weighted probability weighting and value function are displayed in Figure 6.8.

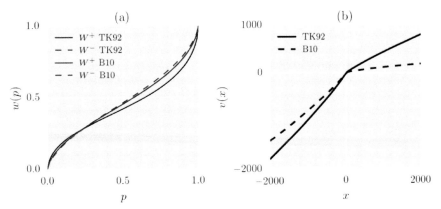

Figure 6.8: (a) Probability weighting function (b) value function TK92 refers to Tversky and Kahneman (1992)and B10 to Booij et al. (2010)

6.3.4   Numerical example

To illustrate the use of the Prospect theory a numerical example is hereby presented where a hypothetical situation is considered. In this case a decision maker is required to decide whether to either *change* the landuse or *not*, under the threat of flooding. Hence, for each decision made and future flooding scenarios may result in either loses or gains that define the prospects.

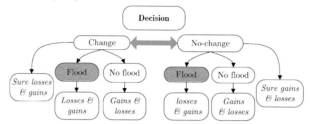

Figure 6.9: Decisions and prospects for the hypothetical example

In this example random probabilities of flooding are used as well as consequences. The consequences are randomly generated from uniform distributions assuming maximum gains and losses up to +100 and -100 units, respectively.

Thus for a hypothetical floodplain location with a probability of flooding of 0.65 (randomly generated) the prospects in Figure 6.9 as randomly generated as:

Table 6-4: Consequence, hypothetical example

State	Consequence (Arbitrary Units)
Gain - Flooded - Change	10.23
Gain - Flooded - No change	88.33
Gain - Not flooded - Change	32.66
Gain - Not flooded - No change	1.99
Sure gain - Change	98.89
Sure gain - No change	80.07
Loss - Flooded - Change	-37.03
Loss- Flooded - No change	-4.98
Loss - Not flooded - Change	-82.38
Loss - Not flooded - No change	-68.48
Sure loss - Change	-44.38
Sure loss - No change	-76.02

Thus the value functions from eq. 6.11 and Figure 6.8(b) is:

$$v(x) = \begin{cases} x^{0.88} & x \geq 0 \\ -2.25 \times (-x)^{0.88} & x < 0 \end{cases}$$

And the weighted probabilities, as per Figure 6.8(a), for gains and losses are calculated as:

$$w^+(p) = \frac{p^{0.61}}{\left(p^{0.61} + (1-p)^{0.61}\right)^{\frac{1}{0.61}}},$$

$$w^-(p) = \frac{p^{0.69}}{\left(p^{0.69} + (1-p)^{0.69}\right)^{\frac{1}{0.69}}}$$

Hence from eq. 6.13 the Cumulative Prospect theory is calculated as:

CU_{Change} =   (gain_sure_Change) + (loss_sure_Change) +

$[v^+$(gain_flooded_Change) $\times w^+$(p) +

v^-(loss_flooded_Change)$\times w^-$(p) +

v^+(gain_not_flooded_Change) $\times w^+$(p) +

v^-(loss_not_flooded_Change)$\times w^-$(1-p)]

CU_{Change} = 98.89 - 44.38 + $\begin{bmatrix} \left(v^+(10.23)\times w^+(0.65)\right) + \left(v^-(-37.03)\times w^-(0.65)\right) + \\ \left(v^+(32.66)\times w^+(1-0.65)\right) + \left(v^-(-82.38)\times w^-(1-0.65)\right) \end{bmatrix}$

= −13.00 *units*

$CU_{No-change}$ =   (gain_sure_No_change) + (loss_sure_No_change) +

$[v^+$(gain_flooded_No_change) $\times w^+$(p) +

v^-(loss_flooded_No_change)$\times w^-$(p) +

v^+(gain_not_flooded_No_change) $\times w^+$(p) +

v^-(loss_not_flooded_No_change)$\times w^-$(1-p)]

$CU_{No-change}$ = 80.07 - 76.02 + $\begin{bmatrix} \left(v^+(88.33)\times w^+(0.65)\right) + \left(v^-(-4.98)\times w^-(0.65)\right) + \\ \left(v^+(1.99)\times w^+(1-0.65)\right) + \left(v^-(-68.48)\times w^-(1-0.65)\right) \end{bmatrix}$

= 65.00 *units*

Therefore $CU_{Change} < CU_{No\ change}$, which implies that the No-change decision for this location is preferable compared to the Change decision, with respect to a flood hazard of a specific magnitude.

6.3.5   Implementation of prospect theory for Ubaye valley case study

Following the calculation of CU (eq. 6.13) either decision a_1 or a_2 (Figure 6.7) with a higher CU value is preferably selected as the better decision. Results of this are presented in Figure 6.10.

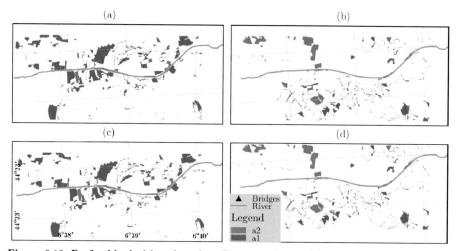

Figure 6.10: Preferable decisions based on Prospect theory (a) TK92 - change (b) TK92 - No change (c) B10 - change (d) B10 - no change

Having chosen two types of stakeholders (Figure 6.8), map locations and corresponding prospect characteristics for which different decisions are identified in Figure 6.11.

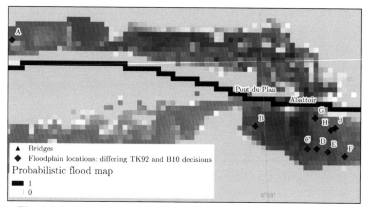

Figure 6.11: Floodplain locations with differing decisions for TK92 and B10 (see Table 6-5)

The data regarding the prospects, landuse, probability of flooding, and decision for the locations in Figure 6.11 are presented in Table 6-5.

Table 6-5: Consequence, probability and decisions for floodplain locations with differing decisions for TK92 refers to Tversky and Kahneman (1992) and B09 refers to Booij et al. (2009) coefficients

	Gains Change	Loss Change	Gains No change	Loss No change	Landuse change: From	Landuse change: To	Probability of flooding	Decision Tk92	Decision B09
A	1152.11	-73.07	0.53	0	fallow	urban	0.78	change	no-change
C	1152.11	-90.91	0.89	-0.31	agriculture	urban	0.85	change	no-change
H	1152.11	-65.85	0.89	-0.31	agriculture	urban	0.78	change	no-change
B	0.53	0	1152.11	-64.83	urban	fallow	0.78	no-change	change
D	0.89	-0.31	1152.11	-106.07	urban	fallow	0.85	no-change	change
E	0.89	-0.31	1152.11	-81.60	urban	fallow	0.82	no-change	change
F	0.89	-0.31	1152.11	-54.83	urban	fallow	0.76	no-change	change
G	0.89	-0.31	1152.11	-56.98	urban	fallow	0.77	no-change	change
J	0.89	-0.31	1152.11	-62.58	urban	fallow	0.78	no-change	change

For these locations that are sensitive to stakeholder characteristics, likelihood of flooding is more certain with likelihood of flooding ranging from 0.77 to 0.85. These locations are also characterised by mainly high investment (fallow to urban landuse change), thus high flood losses. This outcome is intuitive, for which prospects are critical given that flood inundation is highly probable and corresponding consequences (gains and losses) are also high, hence, thus making stakeholder characteristics more determinant when evaluating the decisions to be made.

6.4 CONCLUSION

In this chapter, two approaches have been presented, for which uncertain model outputs (probabilistic flood maps) are used to evaluate landuse change decisions under flood hazard threat. It is a reasonable culmination, having analysed uncertainty in flood inundation modelling and the resulting probabilistic flood maps. These methods can be reproduced with minimal expert knowledge hence making them robust decision making methodologies. Additionally, the methods presented here address challenges represented by situations where information would be necessary to make a decision, and to select between decisions. Similar to the previous chapter (Chapter 5) validation of these methods is currently limited to evaluating the flood damage consequences for past events and observations to evaluate the reliability of the probabilistic map.

The VOI method shows that for cases where there are high distinct consequences and corresponding high uncertainty (likelihood of flooding ≈ 0.5) then high VOI are attained, alluding to further requisite analysis at those locations.

The Prospect theory can be tailored to meet local-needs, characteristics of stakeholders and prevailing conditions. For different types of stakeholders the method yields differing results that are characterised by high alternate consequences are involved.

Following the implementation of the two methodologies, the prospect theory seems to be more intuitive, direct and simpler to implement compared to the Value of information method. With regards, to prospect theory method, flood hazard consequences are evaluated as the actual flood impacts without having to determine a consequence matrix as with the VOI method. The weighted probability is then multiplied by the consequence (that is adapted to cater for loss aversion and diminishing sensitivity), thus this method is apt for evaluating the effects of two or more flood events with varying probabilities of occurrence. The VOI approach is more applicable when components of the modelling process are to be evaluated. For instance, herein, hydraulic model uncertainty is considered for which prior beliefs with respect to flooding is updated to evaluate the VOI. Thus other hydraulic model components uncertainty can be evaluated in this manner. From the outputs of these two methods, the VOI method is more appropriate in supporting spatial planning endeavours by identifying floodplain locations for which extra information should be sought to aid decision makers, while the prospect theory is more suited to evaluate landuse change scenarios. Additionally, the prospect theory approach can be appropriately used (and modified) on different case studies to reflect decision making characteristics and attitudes towards flood risk.

Chapter 7

Conclusions and recommendations

"If I have seen farther than others,
it is because I have stood on shoulders of giants."

Isaac Newton

7.1   INTRODUCTION

Through this thesis, a methodology for inclusion of flood hazard uncertainty in the preventive and mitigation phase of the flood hazard mitigation cycle has been presented. Over the last few decades several authors have worked on the development of methodologies for uncertainty analysis. However, the adoption of these methodologies by stakeholders and decision makers has been limited so far. This is probably due to the potential of compounding ambiguity, or lack of guiding frameworks for the explicit use of uncertain model outputs in flood risk mitigation. As a result, modellers are often (implicitly) pushed to be certain in an uncertain field!

This thesis entails an assessment of several sources of uncertainty that are propagated to generate probabilistic output of the flood inundation modelling, i.e. probabilistic flood maps. These maps do not display flood hazard extent as crisp boundaries, but rather as varying likelihoods of certitude. Therefore, the aim is to avoid providing precisely wrong information, while, on the other hand, being approximately right (Dottori et al. 2013). To assess the usefulness of these maps, two decision making frameworks (value of information and prospect theory) are used. Accordingly, potential landuse changes are evaluated under flood hazard uncertainty that is encapsulated in the probabilistic flood maps. Therefore, resulting from a combination of decision consequences as well as inundation likelihoods, landuse change decisions are evaluated and selected based on higher values of utility. This is done in hind cast, focussing on actual decisions that had been made in the past under the threat of flooding.

Within this chapter the conclusions and recommendations drawn from this study have been presented. These are preceded by a summary of the main results drawn from the study and the limitations associated with the studies undertaken herein.

7.2   SUMMARY OF RESULTS

7.2.1   Uncertainty in flood modelling: Chapter 2 - Chapter 4

Results of hydraulic modelling studies indicate that inflow hydrograph (boundary condition) inaccuracies have a larger effect on the hydraulic model performance when compared to the roughness parameters. Considering the case where a 1D model is assessed while accounting for parametric and rating curve parameter uncertainty; models built with only internal parameters (roughness coefficients) resulted in non-acceptable model results (in terms of model rejection criteria). These models resulted in water level simulations that did not achieve RMSE values less than 1m (rejection criteria herein). On the contrary, models for which rating curve parameters were varied, resulted in several acceptable models (RMSE < 1m), albeit including large spurious RMSE outcomes with values up to 7.0m. Thus, input uncertainties affecting inflow hydrographs and peak values have a significant effect on hydraulic model performance.

The varying action of both rating curve and roughness parameter in a Monte Carlo resulted in prominent identifiability of rating curve parameter and the main channel roughness. Model performance in this case was sensitive to these two parameters that affect the inflow uncertainty and main channel longitudinal flow characteristics. Thus, these models with increased number of parameters had more skill, whereby, several probable states were simulated by the models. This skill can be modified by weighing more accurate ensemble realisations based on actual observations. In addition to the roughness parameters, the inclusion of rating curve uncertainty facilitates the representation of local effects at the upstream boundary to generate and ensemble of inflow hydrographs. Thus these effects are not lumped onto the roughness parameter, more accurate results are yielded.

The 1D model used in this case uses the VDCM (vertical divided channel method) for which there are no momentum interchanges between flows in the main channel and floodplain, thus affecting model performance. Additionally, at the point of embankment overtopping, lateral flows are not simulated resulting in errors in water-level simulation. However, for high flows, shear stress between the main channel and floodplain flow reduces. Hence model performance is sensitive to the roughness parameter. The 1D model in this thesis had only roughness as the free parameter. In this case, all energy losses due to friction and turbulence in addition to local flow effects such as river flow at bends are lumped onto this parameter. Consequently, acceptable performance for the 1D model indicates minimal effects of meanders and complex bathymetry that would lead to significant flow variation along the study reach (Cremona to Borgoforte) along river Po.

An accurate (as possible) understanding of actual flow processes aptly reduces uncertainty due to the selection of an appropriate model structure. Results of the comparison of several sources of uncertainty, model structures, data sources and parameters yielded poor results due to inadequacy of the model structure. Thus, counter-intuitively, concluding that choosing a simple model structure together with coarse resolution dataset yielded reliable results. Though it is emphasised that this outcome is dependent on one case study, hence more testing is required. These results are out of the ordinary given that perceptually it is expected that in reality flood flows are two dimensional in the floodplain thus a 2D model would perform better than a 1D model taking into account the lateral diffusion of the floodwaters. In this case, 1D model performance was better alluding to pre-eminence of 1D flow in for high magnitude events on river Po.

Coarse-resolution datasets are hardly expected to preserve flow controlling line features such as embankments (Figure 4.9). Due to the low resolution of the data, these features are expected to be represented as discontinuous segments in topographical datasets. 2D models built on these datasets were independently conditioned and validated on point water-level measurements; however, more important spatial inundation pattern assessment was not done. In this case there are no historical observed flood inundation extent patterns. The EUDEM performs relatively similarly to SRTM, hence it is concluded that the fusion of ASTER GDEM did not significantly improve utility of the topographic dataset for the case studies herein. On the other hand the ASTER GDEM is reported to have significant noise and thus can also affect the use of the dataset in hydraulic modelling studies.

7.2.2   Usefulness of uncertain information: Chapter 5 and Chapter 6

The second part of the thesis investigates the usefulness of uncertain information (probabilistic flood mapping) in supporting the decision making process in flood hazard mitigation. To this end, value of information and prospect theory are used. In particular, potential consequences are calculated in terms of flood damage. These approaches resulted in the selection of landuse change decisions that yielded higher utilities with respect to potential flood damage consequences. Specifically, application of the prospect theory facilitated the inclusion of actual stakeholders and case study specific characteristics to evaluate decisions to be chosen.

The probabilistic maps contain encapsulated information emanating from sources of uncertainty of the hydraulic model. Hence, this information is found to be robust in supporting the development of flood hazard mitigation strategies, by accounting for several possible outcomes of the model simulations.

These approaches also highlight the challenge in flood risk mitigation where decisions have to be made under uncertainty. High VOI are obtained where there is maximum uncertainty (inundation likelihood $\approx$ 0.5) as well as corresponding distinct large differences between consequences of alternate landuse change decisions. Correspondingly, outcomes for the Prospect theory indicate that where large consequences, involving considerable changes, either to or from urban landuse class result in higher sensitivities of the type of stakeholder due to higher prospects

For the Ubaye Valley case study (France), with respect to actual risk mitigation plans of the case study area, locations with high VOI correspond to locations in the PPR map that require approval prior to landuse alterations. In this case, this is a validation of the procedure.

Results herein support the rational decision to always develop the floodplain, especially for the occurrence of mild - floods. This is preferable given that the welfare trajectory (after the shock of flood damages) remains relatively higher than the alternative option of no-change and a flood occurs.

A comparison of flood hazard mitigation strategies by different types of stakeholders (using the Prospect theory) shows these differences are exaggerated for areas where the likelihood of flooding is high, as well as consequences of landuse change.

Application of the prospect theory necessitates the derivation of weights characterising the stakeholders. Hence, stakeholder and modeller interactions are necessary to achieve this. These interactions ultimately improve awareness of the flood mitigation strategies developed (with respect to spatial planning).

7.3   LIMITATIONS OF THE STUDY

The summary of results presented in the previous section is based on a limited number of case studies and datasets. Limitations of this study are presented in this section.

First, the cases presented herein are based on historical flood events; with flood magnitudes that were approximately similar. Yet, future changes are largely unknown, such as, landuse changes and hydrological regime uncertainties (either increasing or decreasing) among others. Thus, it is expected that predictive uncertainties may be more significant especially when models are used to simulate inputs that are beyond extrapolation zones of data used for conditioning. Scientific studies and discussion regarding future climatic conditions are largely uncertain. Due to expected higher temperatures, increasing hydrological cycle component flux rates would increase with corresponding extremes. This is not accounted for in this thesis.

Anthropogenic activities result in changing landuse patterns that affects surface roughness values. Such conditions are not simulated in this study. In addition, manmade structural defences can alter floodplain inundation thresholds. This was partially demonstrated in the calculation of flood hazard consequences by including channel improvement measures. Run-off generation is heavily dependent on the catchment surface infiltration rates. Increased paved surfaces result in faster runoff generation and catchment response. Surface changes were not tested in this thesis. Though, an assumption was made that during flood flows, (that normatively occur following intense and long duration precipitation seasons), soils are saturated with negligible infiltration during floodplain inundation. It is expected that the shape of the rising limb of the input hydrograph has an effect on flood wave propagation and attenuation. The simulated rising limb is affect by changing roughness, where, as flow volumes increase, energy losses due to friction at the bed decrease. On the contrary, these models are run with a single roughness parameter set for the whole event. Thus, dynamic varying roughness values were not simulated.

Rating curve uncertainty often results in spuriously high peak discharges especially for a single segment rating curve (as used in this thesis), when used in the extrapolation zone. Though in flood inundation modelling worst case scenarios of flooding are of importance, this extrapolation leads to higher peaks, consequently over estimation of flood extents. On the contrary, it could also be argued that water levels that would be catered for by additional segments to the rating curve are relatively few. Thus, more parameters would be generated for which data to constrain them is largely unavailable especially for destructive flood events.

In terms of flood routing models, river bathymetry changes during simulation were assumed to be negligible. This understanding of the river bathymetry was expected for the river Po case study which is characterised by a stable river bed. For Ubaye River, upper catchment forestation and several sediment check-dams along tributary creeks, reduce sediment load deposited in the river. Thus this assumption may have led to errors, though they are expected to be low.

Flood hazard consequence calculation accuracy is dependent upon inputs into the flood damage assessment. Specifically, this mainly includes flood hazard metrics, exposure data (in terms of landuse maps), susceptibility, and value functions. Value functions are largely localised, uncertain and fluctuate temporally. To minimise effects of variations in value functions, assessment is based on a comparative analyses of alternative scenarios and alternate flood mitigation decisions.

The flood hazard in this case is defined as the dual effect of water depth and flow velocity. Different combinations of depth and velocity result in threats to floodplain receptors. Water depth simulation can be validated against observations, however, simulated flow velocity values are potentially erroneous due to lack of data to validate (evaluate) the accuracy. Moreover, within this thesis flood damage was assumed to be a function of the maximum water depth and maximum velocity during the flooding event. A conservative approach was followed by using the maximum values of these two variables. However, it is expected that the timing of maximum velocity is different from the time of maximum water level.

Landuse maps are primary sources of exposure information. However, if these maps are inaccurate or have a low resolution, damage estimates will be inaccurate. Often, these data are available at regional scales and can be processed from satellite remote sensed images. Advances in GIS technology and accessibility such as crowd-sourced 'OpenStreetMaps', avails vector data of receptors (including landuse classes). On the other hand, these data may be used to avail records following verification. Related to exposure maps are susceptibility curves that indicate levels of receptor damage with respect to hazard magnitude. Uncertainties in these functions were not considered, but could significantly alter physical damage thresholds.

Lastly, with regards to decision making within the prospect theory framework, actual stakeholder characteristics related to loss aversion and risk appetite were based on non-flood mitigation studies. These weights are considered as a major factor in the choice of this methodology. Further, studies could be better tailored to characteristics of decision makers in relation to flood mitigation. Such an endeavour can be compared with other regional factors to evaluate stakeholders with respect to decision making under uncertainty.

7.4   CONCLUSIONS

This thesis has assessed the reliability of 2D models of reduced complexity and low order 1D models via calibration, validation and uncertainty analysis. This thesis affirmed the importance of an explicit assessment of uncertainty in flood inundation modelling and the need to portray hydraulic model results in terms of probabilistic maps. The thesis also tested possible methodologies to incorporate this probabilistic information in spatial planning decisions.

Hydraulic model performance is found to vary considerably more due to input hydrograph uncertainty than to model parameter uncertainty. Despite using best practices to measure water levels and velocity (by the relevant river authorities), errors affecting rating curves are bound to generate inaccuracies in input discharge hydrographs.

For Monte Carlo simulations with sampling of rating curve parameters, in addition to the normative hydraulic model parameters resulted in a larger number of good performing models (with respect to the rejection criteria). However, an increase in the number of bad models for this case was also observed.

Reduced complexity flood inundation models simulated historical flood characteristics to within acceptable limits, as compared to recorded observations. Representation of flood flow characteristics (with respect to the properties of the inflow hydrograph) is the most important factor when selecting the appropriate model structure.

The use of Value of Information and Prospect theory, showed that uncertain flood hazard information, encapsulated in probabilistic flood maps, can be incorporated into spatial planning with regards to flood hazard mitigation. The comparison of changing the landuse (or not) resulted in utilities for either of the decisions in the floodplain, hence selecting the option with a higher utility. In addition to selecting preferable decisions, the use of Prospect theory facilitates the inclusion of stakeholder characteristics with regards to their risk nature (i.e. either risk averse or risk seeking) in the analysis.

7.5   RECOMMENDATIONS

We tested the performance of different models of reduced complexity and the results were often satisfactory. Though, given the limited number of case studies and flood events, the outcomes presented here are not conclusive. Thus, further testing of these models on diverse case studies with varying channel characteristics is necessary.

Increasing the number of model parameters (by including rating curve parameters) may result in parameter compensation. This aspect is not considered here. To deal with this, the limits-of-acceptability approach may be used. Additionally, further testing of intermediate complexity models is warranted given the simplification of the models used in this thesis with respect to spatially distributed roughness values.

The use of higher order rating curves is necessitated by high errors in the discharge for a single segment rating curve (that is used in this thesis) especially in the extrapolation zone.

Generally, uncertainty inevitably results in uncertain model outputs that may be presented as probabilistic information. These outputs portray the modeller's level of certitude. Accordingly, these results should be absorbed into decision making frameworks with respect to flood mitigation and spatial planning.

To incorporate flood modelling uncertainty into landuse planning and flood risk mitigation is not limited to the methods presented in this thesis. Other possible frameworks (and methodologies) exist and can be tested. For instance "game theory" can be adapted to evaluate landuse change with respect to probabilities of flooding, thereby yielding dominant implementation strategies.

The framework presented herein is tested on one case study due to data availability limitations and preselected case studies for the project that this work was done under. For future work, in order to generalise the framework further more data-rich case studies maybe used. For instance the uncertainty parameters characterising decision makers herein is based on values derived from literature reviews of studies depicting a large varying population sample size. Thus, parameters may be determined for stakeholders who deal with flood hazard mitigation. The availability of data-rich case studies can facilitate the validation of flood hazard consequences.

A randomised search of the a-priori parameter space is used. However, this may be time consuming especially for computationally demanding higher dimension hydraulic models. Thus more representative and economical sampling techniques such as low-discrepancy sequences (for instance Sobol sequence) and Latin-hypercube sampling methods maybe applied.

Considering temporal hydrological change due to the climate and landuse changes as well as anthropological activities, is also recommended. To this end, the proposed framework can be used to explore how future hydrological scenarios result into different probabilistic maps and this can help evaluating strategies and measures of flood risk reduction.

References

AIPO, 2015. *Monitoraggio Idrografico* [online]. Italy: Interregional Agency for the Po River - AIPO. Available from: http://www.agenziainterregionalepo.it/dati-idrologici.html [Accessed 15th February, 2015 2015].

Alcock, R., Leedal, D. and Beven, K. 2010. Uncertainty and Good Practice in Hydrological Prediction. *VATTEN,* 66, 159-163.

Alfonso, L., Mukolwe, M. M. and Di Baldassarre, G. 2016. Probabilistic flood maps to support decision-making: Mapping the Value of Information. *Water Resources Research,* 52(2).

Alfonso Segura, J. L., 2010. *Optimisation of monitoring networks for water systems: Information theory, value of information and public participation.* TU Delft, Delft University of Technology.

Allsop, W., Kortenhaus, A., Morris, M., Buijs, F., Hassan, R., Young, M., Doorn, N., van der Meer, J., van Gelder, P. and Dyer, M. 2007. Failure mechanisms for flood defence structures. *FLOODsite-Integrated Flood Risk Assessment and Management Methodologies. Res.-Report,* 4.

Angignard, M., 2011. *Applying risk governance principles to natural hazards and risks in mountains.* Dortmund University of Technology.

Apel, H., Merz, B. and Thieken, A. H. 2009. Influence of dike breaches on flood frequency estimation. *Computers & Geosciences,* 35(5), 907-923.

Apel, H., Thieken, A. H., Merz, B. and Blöschl, G. 2004. Flood risk assessment and associated uncertainty. *Natural Hazards and Earth System Sciences,* 4(2), 295-308.

Aronica, G., Bates, P. and Horritt, M. 2002. Assessing the uncertainty in distributed model predictions using observed binary pattern information within GLUE. *Hydrological Processes,* 16(10), 2001-2016.

Aronica, G., Hankin, B. and Beven, K. 1998. Uncertainty and equifinality in calibrating distributed roughness coefficients in a flood propagation model with limited data. *Advances in water resources,* 22(4), 349-365.

Balbi, S., Giupponi, C., Gain, A., Mojtahed, V., Gallina, V., Torresan, S. and Marcomini, A., 2012. *The KULTURisk Framework (KR-FWK): A conceptual framework for comprehensive assessement of risk prevention measures*. Delft, the Netherlands.

Balbi, S., Giupponi, C., Olschewski, R. and Mojtahed, V. 2013. The Economics of Hydro-Meteorological Disasters: Approaching the Estimation of the Total Costs. *BC3 Working Paper Series 2013-12. Basque Centre for Climate Change (BC3), Bilbao, Spain.*

Bates, P. D. 2004. Remote sensing and flood inundation modelling. *Hydrological Processes*, 18(13), 2593-2597.

Bates, P. D., 2005. Flood Routing and Inundation Prediction. *Encyclopedia of Hydrological Sciences*. John Wiley & Sons, Ltd.

Bates, P. D. 2012. Integrating remote sensing data with flood inundation models: how far have we got? *Hydrological Processes*, 26(16), 2515-2521.

Bates, P. D. and De Roo, A. P. J. 2000. A simple raster-based model for flood inundation simulation. *Journal of Hydrology*, 236(1-2), 54-77.

Bates, P. D., Horritt, M. S. and Fewtrell, T. J. 2010. A simple inertial formulation of the shallow water equations for efficient two-dimensional flood inundation modelling. *Journal of Hydrology*, 387(1-2), 33-45.

Bates, P. D., Horritt, M. S., Hunter, N. M., Mason, D. and Cobby, D., 2005. Numerical Modelling of Floodplain Flow. *in Computational Fluid Dynamics: Applications in Environmental Hydraulics (eds P. D. Bates, S. N. Lane and R. I. Ferguson).* Chichester, UK: John Wiley & Sons, Ltd, 271-304.

Beven, K. 2000. Uniqueness of place and process representations in hydrological modelling. *Hydrology and Earth System Sciences*, 4.

Beven, K. 2006. A manifesto for the equifinality thesis. *Journal of Hydrology*, 302(1-2), 18-36.

Beven, K., 2009. *Environmental modelling: an uncertain future?* UK: Routledge.

Beven, K. and Binley, A. 1992. The future of distributed models: model calibration and uncertainty prediction. *Hydrological Processes*, 6(3), 279-298.

Beven, K. and Binley, A. 2013. GLUE: 20 years on. *Hydrological Processes*.

Beven, K., Lamb, R., Leedal, D. and Hunter, N. 2014. Communicating uncertainty in flood inundation mapping: a case study. *International Journal of River Basin Management*, 1-11.

Beven, K. and Westerberg, I. 2011. On red herrings and real herrings: disinformation and information in hydrological inference. *Hydrological Processes,* 25(10), 1676-1680.

Beven, K. J., 2001. *Rainfall-runoff modelling : the primer.* Chichester, UK; New York: J. Wiley.

Bhattacharya, B. and Solomatine, D. P. 2005. Neural networks and M5 model trees in modelling water level–discharge relationship. *Neurocomputing,* 63(0), 381-396.

Boiten, W., 2008. *Hydrometry : a comprehensive indroduction to the measurement of flow in open channels.* Rotterdam, Netherlands: CRC Press, Balkema.

Booij, A., van Praag, B. S. and van de Kuilen, G. 2010. A parametric analysis of prospect theory's functionals for the general population. *Theory and Decision,* 68(1-2), 115-148.

Bouma, J. A., van der Woerd, H. J. and Kuik, O. J. 2009. Assessing the value of information for water quality management in the North Sea. *Journal of Environmental Management,* 90(2), 1280-1288.

Braca, G., 2008. *Stage-discharge relationships in open channels: Practices and problems.* FORALPS Technical Report 11. Universita degli Studi di Trento, Dipartimento di Ingegneria Civile e Ambientale, Trento.

Bracken, L. J., 2013. 7.8 Flood Generation and Flood Waves. *In:* Shroder, J. F. ed. *Treatise on Geomorphology.* San Diego: Academic Press, 85-94.

Brandimarte, L., Brath, A., Castellarin, A. and Baldassarre, G. D. 2009. Isla Hispaniola: A trans-boundary flood risk mitigation plan. *Physics and Chemistry of the Earth, Parts A/B/C,* 34(4–5), 209-218.

Brandimarte, L. and Di Baldassarre, G. 2012. Uncertainty in design flood profiles derived by hydraulic modelling. *Hydrology Research,* 43(6), 753-761.

Breinl, K., Turkington, T. and Stowasser, M. 2013. Stochastic generation of multi-site daily precipitation for applications in risk management. *Journal of Hydrology,* 498(0), 23-35.

Brisson, N., Gate, P., Gouache, D., Charmet, G., Oury, F.-X. and Huard, F. 2010. Why are wheat yields stagnating in Europe? A comprehensive data analysis for France. *Field Crops Research,* 119(1), 201-212.

Brunner, G., 2010. *HEC-RAS, River Analysis System - Hydraulic Reference Manual.* Hydrologic Engineering Center - HEC, Davis CA, USA: US Army Corps of Engineers.

Castellarin, A., Di Baldassarre, G. and Brath, A. 2010. Floodplain management strategies for flood attenuation in the river Po. *River Research and Applications,* 27(8), 1037-1047.

Chow, V. T., 1959. *Open-channel hydraulics : international student ed.* Tokyo etc.: McGraw-Hill Kogakusha.

Citeau, J. 2003. A new flood control concept in the oise catchment area: definition and assessment of flood compatible agricultural activities. *FIG (Fédération Internationale des géometres) Working Week, École Nationale de Sciences Géographiques (ENSG) et Institut de Géographie Nationale,* 1170, 13-17.

Clausen, L. and Clark, P., The development of criteria for predicting dam break flood damages using modelling of historical dam failures. ed. *International conference on river flood hydraulics,* 1990 Chichester, UK, 369-380.

Cobby, D. M., Mason, D. C. and Davenport, I. J. 2001. Image processing of airborne scanning laser altimetry data for improved river flood modelling. *ISPRS Journal of Photogrammetry and Remote Sensing,* 56(2), 121-138.

COMEST, 2005. *The precautionary principle.* Paris, France: World Commission on the Ethics of Scientific Knowledge and Technology (COMEST); United Nations Educational, Scientific and Cultural Organization (UNESCO).

Cunge, J. A., M., H. F. and A., V., 1980. *Practical aspects of computational river hydraulics.* Boston: Pitman Advanced Pub. Program.

De Almeida, G. A. M. and Bates, P. 2013. Applicability of the local inertial approximation of the shallow water equations to flood modeling. *Water Resources Research,* 49(8), 4833-4844.

De Almeida, G. A. M., Bates, P., Freer, J. E. and Souvignet, M. 2012. Improving the stability of a simple formulation of the shallow water equations for 2-D flood modeling. *Water Resour. Res.,* 48(5), W05528.

De Bruijn, K. M. and Klijn, F., Resilient flood risk management strategies. ed. *Presented at: XXIX IAHR CONGRESS* 2001 Beijing, China, 450-457.

de Moel, H., 2012. *Uncertainty in flood risk.* PHD Thesis. VU University Amsterdam.

DEFRA 2006. R&D outputs: Flood risks to people. Phase 2. FD2321/TR1 The flood risks to people methodology. *Department for Environment Food and Rural Affairs and the Environment Agency, London.*

DELTARES, Summary Report. ed. *EGU Topical Meeting: Validation in flood risk Modelling,* 2014 Deflt, Netherlands.

Di Baldassarre, G., 2012. *Floods in a changing climate: Inundation modelling.* Cambridge, UK.: Cambridge University Press.

Di Baldassarre, G. and Claps, P. 2011. A hydraulic study on the applicability of flood rating curves. *Hydrology Research,* 42(1), 10-19.

Di Baldassarre, G., Laio, F. and Montanari, A. 2011. Effect of observation errors on the uncertainty of design floods. *Physics and Chemistry of the Earth, Parts A/B/C,* 42-44, 85-90.

Di Baldassarre, G. and Montanari, A. 2009. Uncertainty in river discharge observations: a quantitative analysis. *Hydrology and Earth System Sciences,* 13(6), 913-921.

Di Baldassarre, G., Schumann, G. and Bates, P. 2009a. Near real time satellite imagery to support and verify timely flood modelling. *Hydrological Processes,* 23(5), 799-803.

Di Baldassarre, G., Schumann, G. and Bates, P. D. 2009b. A technique for the calibration of hydraulic models using uncertain satellite observations of flood extent. *Journal of Hydrology,* 367(3-4), 276-282.

Di Baldassarre, G., Schumann, G., Bates, P. D., Freer, J. E. and Beven, K. J. 2010. Flood-plain mapping: a critical discussion of deterministic and probabilistic approaches. *Hydrological Sciences Journal,* 55(3), 364 - 376.

Di Baldassarre, G. and Uhlenbrook, S. 2012. Is the current flood of data enough? A treatise on research needs for the improvement of flood modelling. *Hydrological Processes,* 26(1), 153-158.

Di Baldassarre, G., Viglione, A., Carr, G., Kuil, L., Salinas, J. L. and Blöschl, G. 2013. Socio-hydrology: conceptualising human-flood interactions. *Hydrology and Earth System Sciences,* 17.

Doll, C. and van Essen, H. 2008. Road infrastructure cost and revenue in Europe. *Produced within the study Internalisation Measures and Policies for all external cost of Transport (IMPACT) P Deliverable,* 2.

Domeneghetti, A., Castellarin, A. and Brath, A. 2012a. Assessing rating-curve uncertainty and its effects on hydraulic model calibration. *Hydrol. Earth Syst. Sci.*, 16(4), 1191-1202.

Domeneghetti, A., Vorogushyn, S., Castellarin, A., Merz, B. and Brath, A. 2012b. Effects of rating-curve uncertainty on probabilistic flood mapping. *Hydrol. Earth Syst. Sci. Discuss.*, 9(8), 9809-9845.

Domeneghetti, A., Vorogushyn, S., Castellarin, A., Merz, B. and Brath, A. 2013. Probabilistic flood hazard mapping: effects of uncertain boundary conditions. *Hydrol. Earth Syst. Sci.*, 17(8), 3127-3140.

Dottori, F., Di Baldassarre, G. and Todini, E. 2013. Detailed data is welcome, but with a pinch of salt: Accuracy, precision, and uncertainty in flood inundation modeling. *Water Resources Research*, 49(9), 6079-6085.

Dottori, F., Martina, M. L. V. and Todini, E. 2009. A dynamic rating curve approach to indirect discharge measurement. *Hydrol. Earth Syst. Sci.*, 13(6), 847-863.

Dottori, F. and Todini, E. 2011. Developments of a flood inundation model based on the cellular automata approach: Testing different methods to improve model performance. *Physics and Chemistry of the Earth, Parts A/B/C*, 36(7-8), 266-280.

Dutta, D., Herath, S. and Musiake, K. 2003. A mathematical model for flood loss estimation. *Journal of Hydrology*, 277(1–2), 24-49.

EU, E. P., 2007. *Directive 2007/60/EC of the European Parliament and of the Council of 23 October 2007 on the assesment and management of flood risks.* Official Journal of the European Union.

European Environmental Agency, 2014. *EU-DEM* [online]. Available from: http://www.eea.europa.eu/data-and-maps/data/eu-dem#tab-metadata [Accessed 23-01-2014 2014].

EXCIMAP, 2007. *Atlas of flood maps : examples from 19 European countries, USA and Japan.* [S.l.] : [s.n.].

Falorni, G., Teles, V., Vivoni, E. R., Bras, R. L. and Amaratunga, K. S. 2005. Analysis and characterization of the vertical accuracy of digital elevation models from the Shuttle Radar Topography Mission. *Journal of Geophysical Research: Earth Surface*, 110(F2), F02005.

Farr, T. G., Rosen, P. A., Caro, E., Crippen, R., Duren, R., Hensley, S., Kobrick, M., Paller, M., Rodriguez, E., Roth, L., Seal, D., Shaffer, S., Shimada, J., Umland,

J., Werner, M., Oskin, M., Burbank, D. and Alsdorf, D. 2007. The Shuttle Radar Topography Mission. *Reviews of Geophysics,* 45(2), RG2004.

Faulkner, H., Parker, D., Green, C. and Beven, K. 2007. Developing a Translational Discourse to Communicate Uncertainty in Flood Risk between Science and the Practitioner. *AMBIO: A Journal of the Human Environment; vol,* 36(8), 692-704.

Fenton, J., Rating curves: part 2, representation and approximation. ed. *6th Conference on Hydraulics in Civil Engineering: The State of Hydraulics,* 2001 Australia, 319-328.

Flageollet, J.-C., Maquaire, O., Martin, B. and Weber, D. 1999. Landslides and climatic conditions in the Barcelonnette and Vars basins (Southern French Alps, France). *Geomorphology; vol,* 30(1-2), 65-78.

Flageollet, J., Maquaire, O. and Weber, D., 1996. *Final National Report.*

FLOODsite, 2007. *Evaluation of Inundation Models: Limits and Capabilities.* Sixth Framework Programme for European Research and Technological Development (2002-2006): European Commision T08-07-01.

Franchini, M., Lamberti, P. and Di Giammarco, P. 1999. Rating curve estimation using local stages, upstream discharge data and a simplified hydraulic model. *Hydrol. Earth Syst. Sci.,* 3(4), 541-548.

Fread, D. L. 1975. Computation of Stage-Discharge Relationships affected by Unsteady Flow. *JAWRA Journal of the American Water Resources Association,* 11(2), 213-228.

Genovese, E. 2006. A methodological approach to land use-based flood damage assessment in urban areas: Prague case study. *Technical EUR Reports, EUR,* 22497.

Giupponi, C., Mojtahed, V., Gain, A. K., Biscaro, C. and Balbi, S., 2015. Chapter 6 - Integrated Risk Assessment of Water-Related Disasters. *In:* John F. Shroder, P. P., Giuliano Di Baldassarre ed. *Hydro-Meteorological Hazards, Risks and Disasters.* Boston: Elsevier, 163-200.

Glenis, V., McGough, A., Kutija, V., Kilsby, C. and Woodman, S. 2013. Flood modelling for cities using Cloud computing. *Journal of Cloud Computing,* 2(1), 1-14.

Götzinger, J. and Bárdossy, A. 2008. Generic error model for calibration and uncertainty estimation of hydrological models. *Water Resour. Res.*, 44, W00B07.

Green, C., Viavattene, C. and Thompson, P., 2011. *Guidance for assessing flood losses.* Middlesex. UK: Flood Hazard Research Centre–Middlesex University.

Gupta, H. V., Beven, K. J. and Wagener, T., 2006. Model Calibration and Uncertainty Estimation. *Encyclopedia of Hydrological Sciences.* John Wiley & Sons, Ltd.

Hall, J. 2004. Comment on 'Of data and models'. *Journal of hydroinformatics*, 6(1), 75-77.

Herschy, R. W., 1999. *Hydrometry : principles and practices.* second ed. Chichester [etc.]: John Wiley & Sons.

Hirshleifer, J. and Riley, J. G. 1979. The analytics of uncertainty and information-an expository survey. *Journal of Economic Literature*, 1375-1421.

Horritt, M. S. 2006. A methodology for the validation of uncertain flood inundation models. *Journal of Hydrology*, 326(1-4), 153-165.

Horritt, M. S. and Bates, P. D. 2002. Evaluation of 1D and 2D numerical models for predicting river flood inundation. *Journal of Hydrology,* 268(1-4), 87-99.

Horritt, M. S., Di Baldassarre, G., Bates, P. D. and Brath, A. 2007. Comparing the performance of a 2-D finite element and a 2-D finite volume model of floodplain inundation using airborne SAR imagery. *Hydrological Processes,* 21(20), 2745-2759.

Hostache, R., Lai, X., Monnier, J. and Puech, C. 2010. Assimilation of spatially distributed water levels into a shallow-water flood model. Part II: Use of a remote sensing image of Mosel River. *Journal of Hydrology*, 390(3–4), 257-268.

Howard, R. A. 1966. Information Value Theory. *Systems Science and Cybernetics, IEEE Transactions on,* 2(1), 22-26.

Howard, R. A. 1968. The Foundations of Decision Analysis. *Systems Science and Cybernetics, IEEE Transactions on,* 4(3), 211-219.

Hunter, N. M., Bates, P. D., Horritt, M. S., de Roo, P. J. and Werner, M. G. F. 2005a. Utility of different data types for calibrating flood inundation models within a GLUE framework. *Hydrology and Earth System Sciences*, 9(4), 412-430.

Hunter, N. M., Bates, P. D., Horritt, M. S. and Wilson, M. D. 2007. Simple spatially-distributed models for predicting flood inundation: A review. *Geomorphology; vol,* 90(3-4), 208-225.

Hunter, N. M., Bates, P. D., Neelz, S., Pender, G., Villanueva, I., Wright, N. G., Liang, D., Falconer, R. A., Lin, B. and Waller, S. 2008. Benchmarking 2D hydraulic models for urban flooding. *Water management; vol,* 161(1), 13-30.

Hunter, N. M., Horritt, M. S., Bates, P. D., Wilson, M. D. and Werner, M. G. F. 2005b. An adaptive time step solution for raster-based storage cell modelling of floodplain inundation. *Advances in water resources,* 28(9), 975-991.

INSEE, 2014. *National institute of statistics and economic studies (Institut national de la statistique et des études économiques)* [online]. Available from: www.insee.fr 2014].

ISO 1996. Measurement of liquid flow in open channels -- Part 1: Establishment and operation of a gauging station.

ISO, 2010. Hydrometry -- Measurement of liquid flow in open channels -- Part 2: Determination of the stage-discharge relationship. 34.

Jansen, P. P., 1979. *Principles of river engineering : the non-tidal alluvial river.* London; San Francisco: . Pitman.

Jones, B. E. 1916. A method of correcting river discharge for a changing stage. *US Geol. Survey Water Supply Paper 375-E.*

Jongman, B., Kreibich, H., Apel, H., Barredo, J. I., Bates, P. D., Feyen, L., Gericke, A., Neal, J., Aerts, J. C. J. H. and Ward, P. J. 2012. Comparative flood damage model assessment: towards a European approach. *Nat. Hazards Earth Syst. Sci.,* 12(12), 3733-3752.

Joslyn, S. L. and LeClerc, J. E. 2012. Uncertainty forecasts improve weather-related decisions and attenuate the effects of forecast error. *J Exp Psychol Appl,* 18(1), 126-140.

Kahneman, D. and Tversky, A. 1979. Prospect Theory: An Analysis of Decision under Risk. *Econometrica,* 47(2), 263-291.

Kayastha, N., 2014. Refining the Committee Approach and Uncertainty Prediction in Hydrological Modelling. Delft, The Netherlands: CRC Press / Balkema.

Kidson, R. and Richards, K. 2005. Flood frequency analysis: assumptions and alternatives. *Progress in Physical Geography,* 29(3), 392-410.

Klemes, V. 1989. The improbable probabilities of extreme floods and droughts. *Hydrology of disasters, James and James, London*, 43-51.

Klijn, F., van Buuren, M. and van Rooij, S. A. 2004. Flood-risk management strategies for an uncertain future: living with Rhine river floods in the Netherlands? *AMBIO: A Journal of the Human Environment*, 33(3), 141-147.

Koutsoyiannis, D. 2015. Generic and parsimonious stochastic modelling for hydrology and beyond. *Hydrological Sciences Journal*, null-null.

Kreibich, H., van den Bergh, J. C. J. M., Bouwer, L. M., Bubeck, P., Ciavola, P., Green, C., Hallegatte, S., Logar, I., Meyer, V., Schwarze, R. and Thieken, A. H. 2014. Costing natural hazards. *Nature Clim. Change*, 4(5), 303-306.

Krzysztofowicz, R. 2001. The case for probabilistic forecasting in hydrology. *Journal of Hydrology*, 249(1-4), 2-9.

Kundzewicz, Z. W., Kanae, S., Seneviratne, S. I., Handmer, J., Nicholls, N., Peduzzi, P., Mechler, R., Bouwer, L. M., Arnell, N., Mach, K., Muir-Wood, R., Brakenridge, G. R., Kron, W., Benito, G., Honda, Y., Takahashi, K. and Sherstyukov, B. 2014. Flood risk and climate change: global and regional perspectives. *Hydrological Sciences Journal*, 59.

Langley, R. 2000. Unified Approach to Probabilistic and Possibilistic Analysis of Uncertain Systems. *Journal of Engineering Mechanics*, 126(11), 1163-1172.

Lavalle, C., Barredo, J., McCormick, N., Engelen, G., White, R. and Uljee, I., 2004. *The MOLAND model for urban and regional growth forecast - A tool for the definition of sustainable development paths*. Ispra, Italy: European Commision: Joint Research Centre.

Lecarpentier, C., 1963. *La crue de juin 1957 et ses conséquences morphodynamiques*. (Thèse de Doctorat; Centre de Géographie Appliquée, Faculté des Lettres et des Sciences Humaines). Universite De Strasbourg, Strasbourg, France.

Leedal, D., Neal, J., Beven, K., Young, P. and Bates, P. 2010. Visualization approaches for communicating real-time flood forecasting level and inundation information. *Journal of Flood Risk Management*, 3(2), 140-150.

LeFavour, G. and Alsdorf, D. 2005. Water slope and discharge in the Amazon River estimated using the shuttle radar topography mission digital elevation model. *Geophysical Research Letters*, 32(17), L17404.

Loat, R. and Petrascheck, A., 1997. *Consideration of Flood Hazards for Activities with Spatial Impact. Recommendations*. Bern, Switzerland: EDMZ.

Lumbroso, D., Asselman, N. E. M., Bakonyi, P., Gaume, E., Logtmeijer, C., Nobis, A. and Woods-Ballard, B., 2007. *Review report of operational flood management methods and models.* : FLOODsite Project Report ; WL Delft Hydraulics.

Marchi, E., Roth, G. and Siccardi, F., The November 1994 Flood Event on the Po River: Structural and Non-Structural Measures Against Inundations. ed. *U.S. - Italy Research Workshop on the Hydrometeorology, Impact, and Management of Extreme Floods.*, 1995.

Marchi, E., Roth, G. and Siccardi, F. 1996. The Po: Centuries of river training. *Physics and Chemistry of The Earth*, 20(5-6), 475-478.

Masoero, A., Claps, P., Asselman, N. E. M., Mosselman, E. and Di Baldassarre, G. 2013. Reconstruction and analysis of the Po River inundation of 1951. *Hydrological Processes*, 27(9), 1341-1348.

Mason, D. C., Bates, P. D. and Dall' Amico, J. T. 2009. Calibration of uncertain flood inundation models using remotely sensed water levels. *Journal of Hydrology*, 368(1–4), 224-236.

Mason, D. C., Cobby, D. M., Horritt, M. S. and Bates, P. D. 2003. Floodplain friction parameterization in two-dimensional river flood models using vegetation heights derived from airborne scanning laser altimetry. *Hydrological Processes*, 17(9), 1711-1732.

McCarthy, S., Tunstall, S., Parker, D., Faulkner, H. and Howe, J. 2007. Risk communication in emergency response to a simulated extreme flood. *Environmental Hazards*, 7(3), 179-192.

McMillan, H. K. and Brasington, J. 2007. Reduced complexity strategies for modelling urban floodplain inundation. *Geomorphology*, 90(3–4), 226-243.

McMillan, H. K. and Brasington, J. 2008. End-to-end flood risk assessment: A coupled model cascade with uncertainty estimation. *Water Resources Research; vol*, 44(3), W03419.

Md Ali, A., Solomatine, D. P. and Di Baldassarre, G. 2015. Assessing the impact of different sources of topographic data on 1-D hydraulic modelling of floods. *Hydrol. Earth Syst. Sci.*, 19(1), 631-643.

Mersel, M. K., Smith, L. C., Andreadis, K. M. and Durand, M. T. 2013. Estimation of river depth from remotely sensed hydraulic relationships. *Water Resources Research*, 49, 3165-3179.

Merwade, V., Olivera, F., Arabi, M. and Edleman, S. 2008. Uncertainty in Flood Inundation Mapping: Current Issues and Future Directions. *Journal of Hydrologic Engineering*, 13(7), 608.

Merz, B., Thieken, A. H. and Gocht, M., 2007. Flood Risk Mapping At The Local Scale: Concepts and Challenges. *In:* Begum, S., Stive, M. F. and Hall, J. eds. *Flood Risk Management in Europe.* Netherlands: Springer Netherlands, 231-251.

Messner, F., Penning-Rowsell, E., Green, C., Meyer, V., Tunstall, S. and van der Veen, A. 2007. Evaluating flood damages: guidance and recommendations on principles and methods. *FLOODsite-Report T09-06-01.*

Mignot, E., Paquier, A. and Haider, S. 2006. Modeling floods in a dense urban area using 2D shallow water equations. *Journal of Hydrology*, 327(1–2), 186-199.

Montanari, A. 2007. What do we mean by 'uncertainty'? The need for a consistent wording about uncertainty assessment in hydrology. *Hydrological Processes*, 21(6), 841-845.

Mukolwe, M. M., Di Baldassarre, G. and Bogaard, T., 2015a. Chapter 7 - KULTURisk Methodology Application: Ubaye Valley (Barcelonnette, France). *In:* Baldassarre, J. F. S. P. D. ed. *Hydro-Meteorological Hazards, Risks and Disasters.* Boston: Elsevier, 201-211.

Mukolwe, M. M., Di Baldassarre, G., Werner, M. G. F. and Solomatine, D. P. 2014. Flood modelling: parameterisation and inflow uncertainty. *Proceedings of the ICE - Water Management*, 167, 51-60.

Mukolwe, M. M., Yan, K., Di Baldassarre, G. and Solomatine, D. 2015b. Testing new sources of topographic data for flood propagation modelling under structural, parameter and observation uncertainty. *Hydrological Sciences Journal*, 61(9).

Neal, J., Fewtrell, T. and Trigg, M. 2009a. Parallelisation of storage cell flood models using OpenMP. *Environmental Modelling & Software*, 24(7), 872-877.

Neal, J., Keef, C., Bates, P., Beven, K. and Leedal, D. 2013. Probabilistic flood risk mapping including spatial dependence. *Hydrological Processes*, 27(9), 1349-1363.

Neal, J., Schumann, G. and Bates, P. 2012a. A subgrid channel model for simulating river hydraulics and floodplain inundation over large and data sparse areas. *Water Resour. Res.*, 48(11), W11506.

Neal, J. C., Bates, P. D., Fewtrell, T. J., Hunter, N. M., Wilson, M. D. and Horritt, M. S. 2009b. Distributed whole city water level measurements from the Carlisle 2005 urban flood event and comparison with hydraulic model simulations. *Journal of Hydrology*, 368(1-4), 42-55.

Neal, J. C., Villanueva, I., Wright, N., Willis, T., Fewtrell, T. and Bates, P. 2012b. How much physical complexity is needed to model flood inundation? *Hydrological Processes*, 26.

Neilson, W. and Stowe, J. 2002. A Further Examination of Cumulative Prospect Theory Parameterizations. *Journal of Risk and Uncertainty*, 24(1), 31-46.

OECD, 2012. *The value of statistical life: a meta analysis*. Paris, Fr.: Organisation for Economic Co-operation and Development.

Pappenberger, F. and Beven, K. 2006. Ignorance is bliss: Or seven reasons not to use uncertainty analysis. *Water Resources Research.*, 42(5), W05302.

Pappenberger, F., Harvey, H., Beven, K., Hall, J. and Meadowcroft, I. 2006a. Decision tree for choosing an uncertainty analysis methodology: a wiki experiment http://www.floodrisknet.org.uk/methodshttp://www.floodrisk.net. *Hydrological Processes*, 20(17), 3793-3798.

Pappenberger, F., Matgen, P., Beven, K. J., Henry, J.-B., Pfister, L. and Fraipont de, P. 2006b. Influence of uncertain boundary conditions and model structure on flood inundation predictions. *Advances in water resources*, 29(10), 1430-1449.

Parisi, V. R., 2002. Floodplain Management and Mitigation in France. *Disaster Mitigation in Urbanized Areas*. Paris, France: Association of State Floodplain Managers and FEMA Region V.

Patro, S., Chatterjee, C., Singh, R. and Raghuwanshi, N. S. 2009. Hydrodynamic modelling of a large flood-prone river system in India with limited data. *Hydrological Processes*, 23(19), 2774-2791.

Pelletier, P. M. 1988. Uncertainties in the single determination of river discharge: a literature review. *Canadian Journal of Civil Engineering*, 15(5), 834-850.

Perez, F. and Granger, B. E. 2007. IPython: a system for interactive scientific computing. *Computing in Science & Engineering*, 9(3), 21-29.

Popper, W. and Berkeley, W., 1951. *The Cairo Nilometer*. University of California Press.

Preissmann, A., Propagation of translatory waves in channels and rivers. ed. *Proc, First Congress of French Assoc, for Computation, Genoble, France*, 1961, 433-442.

Prestininzi, P., Di Baldassarre, G., Schumann, G. and Bates, P. D. 2011. Selecting the appropriate hydraulic model structure using low-resolution satellite imagery. *Advances in water resources*, 34(1), 38-46.

Ramos, M.-H., Mathevet, T., Thielen, J. and Pappenberger, F. 2010. Communicating uncertainty in hydro-meteorological forecasts: mission impossible? *Meteorological Applications*, 17(2), 223-235.

Refsgaard, J. C., van der Sluijs, J. P., Brown, J. and van der Keur, P. 2006. A framework for dealing with uncertainty due to model structure error. *Advances in water resources*, 29(11), 1586-1597.

Refsgaard, J. C., van der Sluijs, J. P., Højberg, A. L. and Vanrolleghem, P. A. 2007. Uncertainty in the environmental modelling process – A framework and guidance. *Environmental Modelling & Software*, 22(11), 1543-1556.

Reitan, T. and Petersen-Øverleir, A. 2009. Bayesian methods for estimating multi-segment discharge rating curves. *Stochastic Environmental Research and Risk Assessment*, 23(5), 627-642.

Rodriguez, E., Morris, C. S. and Belz, J. E. 2006. A global assessment of the SRTM performance. *Photogrammetric engineering and remote sensing*, 72(3), 249-260.

Romanowicz, R. and Beven, K. 2003. Estimation of flood inundation probabilities as conditioned on event inundation maps. *Water Resources Research*, 39(3), 1073.

Ronco, P., Gallina, V., Torresan, S., Zabeo, A., Semenzin, E., Critto, A. and Marcomini, A. 2014. The KULTURisk Regional Risk Assessment methodology for water-related natural hazards; Part 1: Physical–environmental assessment. *Hydrol. Earth Syst. Sci.*, 18(12), 5399-5414.

RTM, 2006. *Plan de Prevention des Risques Naturels Previsibles- Carte de Zonage Reglementaire: Commune de Barcelonnette*. Haute-Provence, France: Service Departemental de Restauration des Terrains en Montagne des Alpes de Haute-Provence.

RTM, 2009. *Plan De Prevention Des Risques Naturels Previsibles(P.P.R.) - Note De Presentation - Commune De Barcelonnette*. Haute-Provence, France: RTM - Restauration des Terrains en Montagne, No. 2009-2699.

Sanders, B. F. 2007. Evaluation of on-line DEMs for flood inundation modeling. *Advances in water resources*, 30(8), 1831-1843.

Schlichting, H., 1979. *Boundary-layer theory*. 7 ed. New York: McGraw-Hill.

Schumann, G., Bates, P., Horritt, M., Matgen, P. and Pappenberger, F. 2009. Progress in integration of remote sensing–derived flood extent and stage data and hydraulic models. *Reviews of Geophysics*, 47(4), RG4001.

Schwarze, R., Schwindt, M., Weck-Hannemann, H., Raschky, P., Zahn, F. and Wagner, G. G. 2011. Natural hazard insurance in Europe: tailored responses to climate change are needed. *Environmental Policy and Governance*, 21(1), 14-30.

Shrestha, D. L., 2009. *Uncertainty analysis in rainfall-runoff modelling : application of machine learning techniques*. Leiden, The Netherlands: CRC/Balkema.

Smith, L. C. 1997. Satellite remote sensing of river inundation area, stage, and discharge: a review. *Hydrological Processes*, 11(10), 1427-1439.

Solomatine, D. P. and Shrestha, D. L. 2009. A novel method to estimate model uncertainty using machine learning techniques. *Water Resources Research*, 45(12), W00B11.

Starmer, C. 2000. Developments in Non-expected Utility Theory: The Hunt for a Descriptive Theory of Choice under Risk. *Journal of Economic Literature*, 38(2), 332-382.

Stedinger, J. R., Vogel, R. M., Lee, S. U. and Batchelder, R. 2008. Appraisal of the generalized likelihood uncertainty estimation (GLUE) method. *Water Resources Research*, 44(12), W00B06.

Stein, S. and Stein, J., 2014. *Playing Against Nature: Integrating Science and Economics to Mitigate Natural Hazards in an Uncertain World*. John Wiley & Sons.

Straatsma, M., Middelkoop, H. and de Jong, S. 2011. Hydrodynamic roughness of floodplain vegetation: Airborne parameterization and field validation. *LOICZ Research and Studies*, 38, 19-25.

Streiner, D. L. and Norman, G. R. 2006. "Precision" and "Accuracy": Two Terms That Are Neither. *Journal of Clinical Epidemiology*, 59(4), 327-330.

Tachikawa, T., Hato, M., Kaku, M. and Iwasaki, A., Characteristics of ASTER GDEM Version 2. ed. *IEEE International Geoscience and Remote Sensing Symposium*, 2011a Vancouver, Canada.

Tachikawa, T., Kaku, M., Iwasaki, A., Gesch, D., Oimoen, M., Zheng, Z., Danielson, J., Krieger, T., Curtis, B., Haase, J., Abrams, M., Crippen, R. and Carabaja, C., 2011b. *ASTER Global Digital Elevation Model Version 2 - Summary of Validation Report.* Japan.

Thiery, Y., Malet, J. P., Sterlacchini, S., Puissant, A. and Maquaire, O. 2007. Landslide susceptibility assessment by bivariate methods at large scales: Application to a complex mountainous environment. *Geomorphology*, 92(1–2), 38-59.

Tversky, A. and Kahneman, D. 1992. Advances in prospect theory: Cumulative representation of uncertainty. *Journal of Risk and Uncertainty*, 5(4), 297-323.

USACE, 1996. *Engineering and Design; Risk-Based Analysis for Flood Damage Reduction Studies.* Washington DC: U.S. Army Corps of Engineers.

van der Zon, N., 2013. *Kwaliteitsdocument Actueel Hoogtebestand Nederland (AHN2) [www.ahn.nl accessed 28 January 2015].* the Netherlands: HetWaterschapshuis.

Van Gelder, P., 2000. *Statistical methods for the risk-based design of civil structures.* TU Delft, Delft University of Technology.

Vis, M., Klijn, F., De Bruijn, K. and Van Buuren, M. 2003. Resilience strategies for flood risk management in the Netherlands. *International Journal of River Basin Management*, 1(1), 33-40.

Von Neumann, J. and Morgenstern, O., 1953. *Theory of games and economic behavior.* Princeton: Princeton University Press.

Wade, S., Ramsbottom, D., Floyd, P., Penning-Rowsell, E. and Surendran, S., Risks to people: Developing new approaches for flood hazard and vulnerability mapping. ed. *40th Defra Flood and Coastal Management Conference*, 2005 UK.

Walker, W. E., Harremoës, P., Rotmans, J., van der Sluijs, J. P., van Asselt, M. B. A., Janssen, P. and von Krauss, M. P. K. 2003. Defining Uncertainty: A Conceptual Basis for Uncertainty Management in Model-Based Decision Support. *Integrated Assessment*, 4(1), 5 - 17.

Wang, W., Yang, X. and Yao, T. 2012. Evaluation of ASTER GDEM and SRTM and their suitability in hydraulic modelling of a glacial lake outburst flood in southeast Tibet. *Hydrological Processes*, 26(2), 213-225.

Werner, M. G. F., 2004. *Spatial flood extent modelling. A performance based comparison.* Doctoral Thesis. Delft University Press.

Werner, M. G. F., Hunter, N. M. and Bates, P. D. 2005. Identifiability of distributed floodplain roughness values in flood extent estimation. *Journal of Hydrology,* 314(1–4), 139-157.

Winsemius, H. C., Aerts, J. C. J. H., van Beek, L. P. H., Bierkens, M. F. P., Bouwman, A., Jongman, B., Kwadijk, J. C. J., Ligtvoet, W., Lucas, P. L., van Vuuren, D. P. and Ward, P. J. 2015. Global drivers of future river flood risk. *Nature Clim. Change,* advance online publication.

Winsemius, H. C., Van Beek, L. P. H., Jongman, B., Ward, P. J. and Bouwman, A. 2013. A framework for global river flood risk assessments. *Hydrology and Earth System Sciences,* 17, 1871-1892.

Yan, K., Di Baldassarre, G. and Solomatine, D. P. 2013. Exploring the potential of SRTM topographic data for flood inundation modelling under uncertainty. *Journal of hydroinformatics,* 15(3), 849-861.

Yan, K., Di Baldassarre, G., Solomatine, D. P. and Schumann, G. J. P. 2015a. A review of low-cost space-borne data for flood modelling: topography, flood extent and water level. *Hydrological Processes,* n/a-n/a.

Yan, K., Tarpanelli, A., Balint, G., Moramarco, T. and Di Baldassarre, G. 2015b. Exploring the Potential of SRTM Topography and Radar Altimetry to Support Flood Propagation Modeling: Danube Case Study. *Journal of Hydrologic Engineering,* 20(2), 04014048.

Appendix

ACRONYMS

AIPO	*Agenzia Interregionale per il fiume Po* (Interregional Agency for the River Po)
ERRA	Economic - Regional Risk Analysis
ERRA	Socio-Economic - Regional Risk Analysis
PPR	*Plans de Prévention des Risques* (Risk Prevention Plans)
RRA	Regional Risk Analysis
SRRA	Social - Regional Risk Analysis
DEM	Digital Elevation Model
DSM	Digital Surface Model
SRTM	Shuttle Radar Topography Mission
LiDAR	Light Detection and Ranging
EUDEM	European Union Digital Elevation Model
ASTER	Advanced Space borne Thermal Emission and Reflection Radiometer
GDEM	Global Digital Elevation Model
VOI	Value of Information
GLUE	Generalised Likelihood Uncertainty Estimation
RMSE	Root Mean Square Error
VOI	Value of Information
TK92	Tversky and Kahneman (1992)
B10	Booij et al. (2010)
INSEE	*Institut national de la statistique et des études économiques* (National Institute of statistics and economic studies)
PPRI	*Plans de Prevention du Risqué d'Inondations* (Flood Inundation Risk Prevention Plans)
CCR	*Caisse Centrale de Reassuarance*
AIPO	*Agenzia Interregionale per il fiume Po* (Interregional Agency for the River Po)

Samenvatting

Overstromingen zijn natuurlijke gebeurtenissen die kwetsbare samenlevingen kunnen treffen en aanzienlijke schade kunnen veroorzaken. Floodplain mapping, het in kaart brengen van de gebieden die het risico lopen overstroomd te worden, kan helpen om de negatieve gevolgen van overstromingen te reduceren, door het ondersteunen van het opstellen van bestemmingsplannen voor gebieden die blootgesteld zijn aan overstromingsgevaar.

De recente literatuur laat zien dat hydraulische modellering van overstromingen wordt beïnvloed door talloze onzekere factoren die gereduceerd (maar niet geëlimineerd) kunnen worden door kalibratie en validatie. Veel studies hebben bijvoorbeeld aangetoond er niet in te slagen om overstromingen te simuleren die van een andere grootte zijn dan de calibratie en validatie gegevens. Dit kan veroorzaakt worden door het feit dat de stromingsmechanismes van de rivier niet-lineair zijn en gekarakteriseerd worden door drempels die verschillende vloeiregimes afbakenen.

Eén van de uitdagingen bij het gebruik van onzekere resultaten is dat besluitvormers (bv voor ruimtelijke ordening) vaak binaire acties moeten ondernemen, bijvoorbeeld, het bestemmingsplan (bv bebouwing) veranderen of niet. Vanuit het perspectief van een modelleur kunnen precieze (maar mogelijk foute) resultaten aangeleverd worden, gebaseerd zowel op expertise als de resultaten van gekalibreerde en gevalideerde modellen. Echter, dit is noch verstandig noch praktisch, aangezien expertise veranderlijk is en onvermijdelijk subjectief. Het blijkt dat verschillende modelleurs die dezelfde input data en modellen gebruiken, vaak verschillende resultaten krijgen. Het is dus wetenschappelijk gezien beter om de resultaten van overstromingsmodellen in termen van kansberekening weer te geven.

Het doel van dit proefschrift is bij te dragen aan het wetenschappelijke werk om de onzekerheid van overstromingsmodellen in te schatten, en methodes te ontwikkelen om het gebruik van kaarten met overstromingskansen bij de ruimtelijke ordening te bevorderen. De invloed van diverse belangrijke bronnen van onzekerheid (zoals de invoer van de overstromings-hydrografie, model-parameters en structuur) worden ingeschat door te focussen op minder complexe modellen van overstromingsdynamiek. Vervolgens worden nieuwe methodes getest om onzekere uitkomsten van modellen mee te nemen bij besluiten over ruimtelijke ordening in overstromingsgebieden. Meer specifiek, het proefschrift bestaat uit twee (elkaar aanvullende) delen.

Het eerste deel gaat over de analyse van de belangrijkste bronnen van fouten bij het modelleren van overstromingen, wat culmineert in de vervaardiging van kaarten met overstromingskansen. Het tweede deel laat de toepassing zien van bruikbare benaderingen om het besluitvormingsproces te ondersteunen.

Dit proefschrift levert een bijdrage aan het gebruik van kaarten met overstromingskansen bij besluitvorming, zoals bijvoorbeeld ruimtelijke ordening bij onzekerheid rondom overstromingsgevaar. Met historische hydrologische gegevens worden 1D, 1D-2D en 2D overstromingsmodellen gebruikt om overstroming-scenario's te simuleren. Deze modellen zijn gemaakt voor twee case studies: (i) een bergachtig rivierengebied (De rivier de Ubaye, Frankrijk) en (ii) een alluviaal rivierengebied (de rivier de Po, Italië). Topografische data worden ontleend aan veel gebruikte informatiebronnen van verschillende precisie en nauwkeurigheid, namelijk SRTM (Shuttle Radar Topography Mission), EUDEM (Digital Elevation Model over Europe) and LiDAR (Light Detection and Ranging). Er worden met name vier grote bronnen van onzekerheid, die de uitkomsten van overstromingsmodellen beïnvloeden, geanalyseerd. Dat zijn onzekerheid over de toevoer (overstromingstoevloed afgeleid van een getallencurve), over de parameters, de modelstructuur en de topografische data. Onzekerheid over de toevoer werd op twee manieren gedefinieerd: (i) onzekerheid over de "single segment rating curve parameter", en (ii) onzekerheid over onderdelen van de "aggregated peak discharge". De onzekerheid over de grenswaarde (inflow hydrograph) bleek veel significanter dan de parametrische onzekerheid. Kaarten met overstromingskansen worden gegenereerd met een Monte Carlo benadering om de invloed van deze onzekerheden vast te leggen. Tenslotte is een nieuwe methodologie gebruikt om de invloed van maatregelen tegen overstromingsgevaar in te schatten. (het KULTURisk framework , ontwikkeld bij een EU FP7 project)

De bruikbaarheid van probabilistische model output wordt dan ingeschat door twee benaderingen: (i) Waarde van de informatie, en (ii) Vooruitzicht theorie. Implementatie van deze twee benaderingen is gebaseerd op de premisse van een welvaartstraject, waarbij de waarde van bezittingen en investeringen in het overstromingsgebied toenemen in de loop van de tijd. Dus, het optreden van een overstroming resulteert in schade die het welvaarttraject verlagen. Gebruik van het land in het overstromingsgebied kan veranderd worden gebaseerd op de behoeften van de gemeenschap, en tevens op het potentiële overstromingsgevaar. In dit geval levert een grotere investering meer opbrengsten, en dat zorgt voor een hoger welvaartstraject. Een combinatie van winst door verandering van grondgebruik en een corresponderende dreiging van overstromingsschade (gebaseerd op kaarten met overstromingskansen) laat het ruimtelijke ordenings-dilemma zien waar veel besluitvormers mee te maken hebben. In deze context heeft dit proefschrift laten zien dat uitkomsten van probabilistische modellen succesvol gebruikt kunnen worden om strategieën tegen overstromingsgevaar te ontwikkelen, en ruimtelijke ordening te ondersteunen in overstromingsgebieden. Resultaten wijzen ook op reële uitdagingen bij ruimtelijke ordening waar overstromingsgebieden met grotere consequenties en onzekerheid worden geïdentificeerd, zodat ze beter kunnen worden gecontroleerd.

ACKNOWLEDGEMENT

Several people, contributing to different spheres around my life have aided the writing of this script. To thank them all individually is impossible; I only hope they know that I am grateful.

I must mention my friends Benson Magembe and Emmanuel Osilo for their steady patience and encouragement that helped me to bridge the gap between tertiary education and employment. Eng. Dr. Rugumayo has been a consistent mentor giving advice along the way to this point. Eng. M.K. Mukangula and Prof. S.K. Makhanu for having the faith to put me in charge of engineering projects and encouragement to pursue further studies. My M.Sc. course coordinator, Dr. Luigia Brandimarte, thank you for your advice and prompting that pushed me through my studies. To these people I must say thank you, my postgraduate studies may not have been possible without your input.

To Prof. Di Baldassarre my supervisor and initiator of the project that funded my PhD, words fail me to express my gratitude for the opportunity to achieve this. I will always remember the chat that made me believe that I could actually work towards a PhD. I must also thank my promotor Prof. D.P. Solomatine whose critical thinking has improved my scientific outlook and approach to problem solving. Thank you for the independence of thought and memorable PhD journey. I would like to extend my gratitude to Dr. Leonardo Alfonso and Dr. Thom Bogaard whose help and support was instrumental to the success of the KULTURisk project.

To those that helped me smooth-out the highs and lows of PhD life by chatting, sharing, running; thank you Norbert, Ruud, Henk, Maurizio, Girma, Yan, Anuar, Elena, Nagendra, Adrian, Juan, Yared, Kaycee, Janet, Peter, Joel, Eva and Isnaeni. Thanks to all the Kenyan colleagues whom I have met during my stay in Delft over the years, I honestly cannot list you all here.

I feel guilty; and can only hope that I can repay time that I have robbed my beloved daughter, Anna. Thank you Caroline, for the daily encouragement, understanding and support; this far we are and hope the furthest too shall be a reality.

To my Late father Mr. Moses M. Mukolwe and mother Mrs Fredah K. Mukolwe thank you for the endless love and support. Some stopped along the way, but dreams that were are continually unfolding into reality.

Micah M. Mukolwe

Delft, 2016

ABOUT THE AUTHOR

Micah Mukungu Mukolwe was born in Nairobi, Kenya in 1980. He graduated from Makerere University in 2005 with a degree in Civil Engineering. From 2007 to 2009 Micah worked as a graduate Civil Engineer at a young university in Western Kenya, concentrating on civil and structural works appertaining to construction of facilities within the university. Having worked in a rural setting, he gained an appreciation for the ubiquitous water resource availability, quality, and accessibility challenges. In 2009 he joined UNESCO-IHE Master of Science programme (M.Sc.); Water Science and Engineering, Hydraulic engineering and River basin development, having been awarded a scholarship by the Netherlands Fellowship Programme. He graduated with Master of Science degree in 2011. His M.Sc. research in flood inundation modelling was the foundation for his PhD. After his M.Sc. study he began a fulltime PhD research titled: "Flood Hazard Mapping: Uncertainty and its Value in the Decision-making Process" while attached to an EU FP7 project (KULTURisk).

LIST OF PUBLICATIONS

Alfonso, L., Mukolwe, M. M. and Di Baldassarre, G. 2016. Probabilistic flood maps to support decision-making: Mapping the Value of Information. Water Resources Research. 52(2). doi:10.1002/2015WR017378

Mukolwe, M. M., Di Baldassarre, G. and Bogaard, T., 2015a. Chapter 7 - KULTURisk Methodology Application: Ubaye Valley (Barcelonnette, France). In: Baldassarre, J. F. S. P. D. ed. Hydro-Meteorological Hazards, Risks and Disasters. Boston: Elsevier, 201-211.

Mukolwe, M. M., Di Baldassarre, G., Werner, M. G. F. and Solomatine, D. P. 2014. Flood modelling: parameterisation and inflow uncertainty. Proceedings of the ICE - Water Management, 167, 51-60.

Mukolwe, M. M., Yan, K., Di Baldassarre, G. and Solomatine, D. 2015b. Testing new sources of topographic data for flood propagation modelling under structural, parameter and observation uncertainty. Hydrological Sciences Journal, 61(9).

Conference Contributions

Mukolwe M. M., Di Baldassarre G., Werner M., Solomatine D. P. 2013. Boundary condition and parameter uncertainty in 1D hydraulic models. 5th EGU Leonardo conference (Facets of Uncertainty), Kos Island, Greece.

Mukolwe, M. M., Di Baldassarre, G. and Solomatine, D. 2013. The use of sediment deposition maps as auxiliary data for hydraulic model calibration. EGU General Assembly Conference Abstracts - 10367.

Mukolwe, M. M., Di Baldassarre, G., Bogaard, T., Malet, J.-P. and Solomatine, D. 2012. Probabilistic Flood Mapping and Visualization Issues: Application to the River Ubaye, Barcelonnette (France). EGU General Assembly Conference Abstracts - 10871.

Mukolwe M. M., Di Baldassarre G., Solomatine D. P. 2012. Uncertain flood maps as an aid to stakeholder participation in flood modelling. EGU Leonardo conference 2012 (Hydrology and Society). Turin, Italy.

Mukolwe, M. M., Di Baldassarre, G., Werner, M. and Solomatine, D. 2011. Uncertainty in 1D Hydraulic Modelling Caused By Parameterization And Inflow Inaccuracy. Floods in 3D: Processes, Patterns, Predictions, Bratislava, Slovakia.

Alfonso L., Mukolwe M. M., Di Baldassarre G. 2015. Spatial planning using probabilistic flood maps. EGU General Assembly Conference Abstracts - 10681.